AF588066

General Physics II

Mehdi Rahmani-Andebili

General Physics II

Practice Problems, Methods, and Solutions

Springer

Mehdi Rahmani-Andebili
Electrical and Computer Engineering
The University of Alabama
Tuscaloosa, AL, USA

ISBN 978-3-031-92865-9   ISBN 978-3-031-92866-6 (eBook)
https://doi.org/10.1007/978-3-031-92866-6

This Springer imprint is published by the registered company Springer Nature Switzerland AG
The registered company address is: Gewerbestrasse 11, 6330 Cham, Switzerland

Preface

General Physics I and II are two mandatory courses for all science and engineering majors and even for some non-engineering majors that are taught at universities and colleges worldwide for freshman students. This textbook, as the General Physics II, has been prepared for instructors as well as for students taking General Physics courses. In each chapter of the textbook, different types of problems and exercises have been presented that are categorized as follows.

- ***Problems with detailed solution***: They have been designed to teach students the subjects in detail. Moreover, they have been categorized in different levels based on their difficulty levels (easy, normal, and hard) and calculation amounts (small, normal, and large). This classification helps students study the book in the most efficient way.
- ***Partially solved exercises***: They have been designed to encourage students to practice problems while guiding them through the problem-solving procedure and hinting the required formulas.
- ***Exercises with final answer***: They have been designed to encourage students to practice more by themselves while hinting them by the final answer as well as to help instructors to give tests or quizzes.

In the following, the description of each chapter of the textbook is briefly presented.

Chapters 1 and 2 cover the subjects concerned with the Electric Charge, Electric Force, Electric Field, Electric Potential, Electric Potential Energy, Electric Energy, and Electric Energy Density.

Chapters 3 and 4 teach the Capacitance and Equivalent Capacitance of a Circuit, Electrical Charge and Voltage of Capacitors, Electrical Energy and Energy Density Stored in Capacitors, Resistance and Equivalent Resistance of a Circuit, Ohm's Law and Electrical Current and Voltage of Resistors, and Electrical Power and Energy Wasted in Resistors.

Chapters 5 and 6 study the Magnetic Field Intensity Around a Long Current-Carrying Wire, Magnetic Field Intensity at the Center of a Circle and a Circular Arc, Magnetic Energy and Magnetic Energy Density, and Magnetic Force.

Chapters 7 and 8 review the Magnetic Flux, Voltage Induced in a Coil, Voltage Induced in a Wire, Inductance and Equivalent Inductance of a Circuit, and Magnetic Energy Density and Magnetic Energy Stored in Inductors.

Chapters 9 and 10 investigate the Density, Pressure in Fluids, Archimedes' Principle, and Continuity Principle of Fluids.

Chapters 11 and 12 are concerned with the Temperature and Thermometers, Thermal Expansion, Heat Transfer, Laws of Thermodynamics, Entropy and Change of Entropy, and Carnot Machine.

Chapters 13 and 14 are related to the Transverse Waves, Longitudinal Waves, Sound Intensity and Sound Intensity Level, and Doppler Effect.

Chapters 15 and 16 cover the subjects including Reflection and Refraction, Flat Mirror, Concave Mirror, Convex Mirror, Convex Lens, and Concave Lens.

Moreover, the subjects of the General Physics II are as follows.

- *Vectors and coordinate systems*
- *Linear kinematics*
- *Linear dynamics*
- *Collision and centre of mass*
- *Rotational kinematics and dynamics*
- *Simple harmonic motion*

Since the textbook includes the basic and advanced problems with very detailed problem solutions, it can be used as a practicing study guide by students and as a supplementary teaching source by instructors. Moreover, since the problems and exercises have very detailed solutions, the textbook is helpful for under-prepared students. In addition, it is beneficial for knowledgeable students because it includes advanced problems and exercises.

In preparing the problems and solutions, care has been taken to use methods typically found in the primary instructor-recommended textbooks. By considering this key point, the textbook is in the direction of instructors' lectures, and the instructors will not see any untaught and unusual problem solutions in their students' answer sheets.

Tuscaloosa, AL

Mehdi Rahmani-Andebili

The Other Works Published by the Author

AC Electric Machines: Practice Problems, Methods, and Solutions

AC Electrical Circuit Analysis: Practice Problems, Methods, and Solutions, Second Edition

Advanced Electrical Circuit Analysis: Practice Problems, Methods, and Solutions, Second Edition

Applications of Artificial Intelligence in Planning and Operation of Smart Grid

Applications of Fuzzy Logic in Planning and Operation of Smart Grids

Calculus I: Practice Problems, Methods, and Solutions, Second Edition

Calculus II: Practice Problems, Methods, and Solutions

Calculus III: Practice Problems, Methods, and Solutions

DC Electric Machines, Electromechanical Energy Conversion Principles, and Magnetic Circuit Analysis: Practice Problems, Methods, and Solutions

DC Electrical Circuit Analysis: Practice Problems, Methods, and Solutions, Second Edition

Design, Control, and Operation of Microgrids in Smart Grids

Differential Equations: Practice Problems, Methods, and Solutions

Electromagnetics: Practice Problems, Methods, and Solutions

Feedback Control Systems Analysis and Design: Practice Problems, Methods, and Solutions

General Physics I: Practice Problems, Methods, and Solutions

Mathematics of Engineering and Science: Comprehensive Lessons

MATLAB Lessons, Examples, and Exercises: A Tutorial for Beginners and Experts

Planning and Operation of Electric Vehicles in Smart Grid

Planning and Operation of Plug-in Electric Vehicles: Technical, Geographical, and Social Aspects

Power System Analysis: Comprehensive Lessons

Power System Analysis: Practice Problems, Methods, and Solutions

Precalculus: Practice Problems, Methods, and Solutions, Second Edition

Operation of Smart Homes

Contents

About the Author

Mehdi Rahmani-Andebili is an Assistant Professor with the Department of Electrical and Computer Engineering at the University of Alabama. He received his first M.Sc. and Ph.D. degrees in Electrical Engineering from Tarbiat Modares University and Clemson University in 2011 and 2016, respectively, and his second Ph.D. (2nd M.Sc.) degree in Physics and Astronomy from the University of Alabama in Huntsville in 2019. Moreover, he was a Postdoctoral Fellow at the Sharif University of Technology during 2016–2017. As a professor, he has taught a wide selection of courses and labs including Power System Analysis, Electric Machines, Feedback Control Systems Analysis and Design, Signals and Systems, Electromagnetics, Electronics, Digital Logic, Industrial Electronics, Renewable Distributed Generation and Storage, AC Electrical Circuits Analysis, DC Electrical Circuits Analysis, Electrical Circuits and Devices, Fundamental of Electrical Engineering, Essentials of Electrical Engineering Technology, and Algebra and Calculus-Based Physics. Dr. Rahmani-Andebili has over 300 single-author or first-author publications including journal papers, conference papers, textbooks, books, and book chapters that are about the fields of Electrical Engineering, Power Engineering, Physics, Mathematics, and Computer Programming. He is an IEEE Senior Member and a permanent reviewer of many reputable journals. His research areas include Smart Grid, Applications of Artificial Intelligence in Planning and Operation of Power Systems, Integration of Renewables and Energy Storages into Power Systems, Energy Scheduling and Demand-Side Management, Electric Vehicles and Distributed Generation, and Advanced Optimization Techniques in Power System Studies.

Electrostatics: Part A

1

Abstract

In this chapter, the basic and advanced problems of Electrostatics are studied. The subjects include Electric Charge, Electric Force, Electric Field, Electric Potential, Electric Potential Energy, Electric Energy, and Electric Energy Density. Herein, different types of problems and exercises are presented that are categorized as follows.

- ***Problems with detailed solution***: They have been designed to teach students the subjects in detail. Moreover, they have been categorized in different levels based on their difficulty levels (easy, normal, and hard) and calculation amounts (small, normal, and large).
- ***Partially solved exercises***: They have been designed to encourage students to practice problems while guiding them through the problem-solving procedure and hinting the required formulas.
- ***Exercises with final answer***: They have been designed to encourage students to practice more by themselves while hinting them by the final answer as well as to help instructors to give tests or quizzes.

1.1 Electric Charge

Problem

1.1. Determine the number of electrons that form four Coulombs electric charge [1–5]? The magnitude of electric charge of an electron is $e = 1.6 \times 10^{-19}\ C$.

Difficulty level ● Easy ○ Normal ○ Hard
Calculation amount ● Small ○ Normal ○ Large

1) 1250×10^{16}
2) 3.2×10^{-19}
3) 2500×10^{16}
4) 6.4×10^{-19}

1.2 Electric Force

Problem

1.2. Which one of the following choices shows wrong direction for the electric force exerted on the test charge (Fig. 1.1)?

Difficulty level ● Easy ○ Normal ○ Hard
Calculation amount ● Small ○ Normal ○ Large

M. Rahmani-Andebili, *General Physics II*, https://doi.org/10.1007/978-3-031-92866-6_1

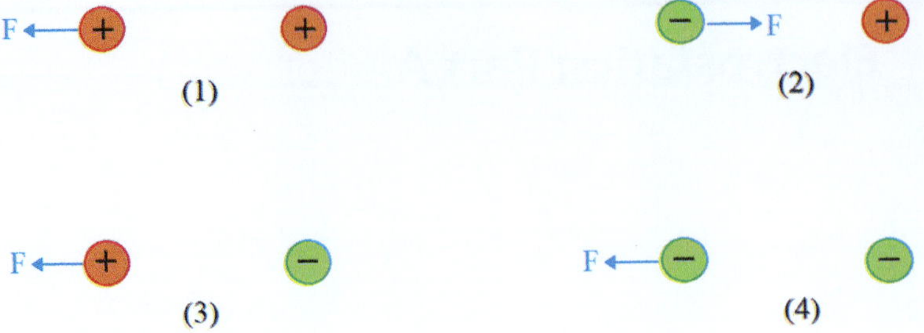

Fig. 1.1 The electric force exerted on the test charge

Problem

1.3. Which one of the following choices is correct about the direction and magnitude of the force that each charge exerts on the other one shown in Fig. 1.2? Herein, $|q_1| > |q_2|$.

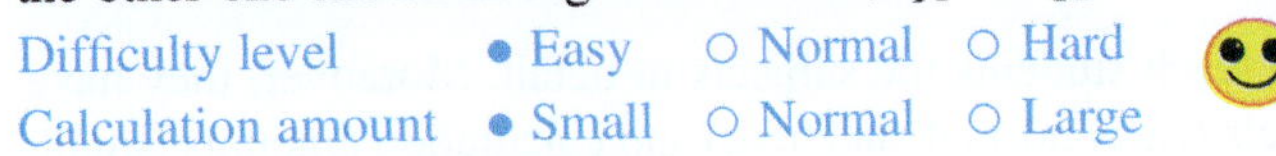

1) $F_{21} = -F_{12}, |F_{21}| > |F_{12}|$
2) $F_{21} = -F_{12}, |F_{21}| < |F_{12}|$
3) $F_{21} = F_{12}, |F_{21}| = |F_{12}|$
4) $F_{21} = -F_{12}, |F_{21}| = |F_{12}|$

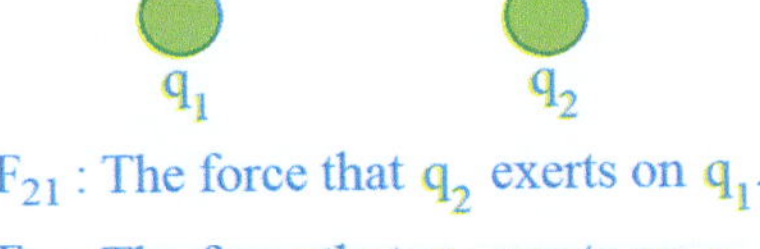

Fig. 1.2 Two charges with unknown types. Herein, $|q_1| > |q_2|$

Problem

1.4. Two electric charges that have been separated from each other exerts an electric force on each other with the magnitude of F. If the magnitude of the charges and their distance are doubled, the magnitude of the force becomes F'. Calculate the value of $\frac{F'}{F}$.

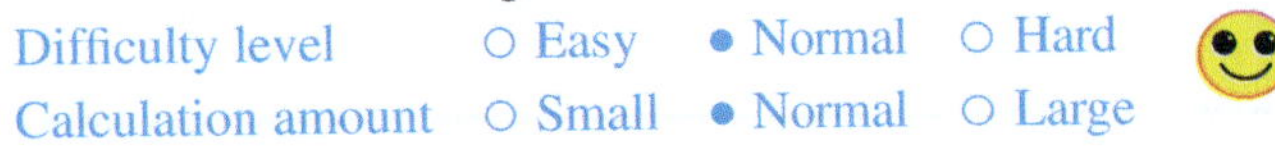

1) 8
2) 2
3) $\frac{1}{2}$
4) 1

Partially Solved Exercise

1.5. Two electric charges that have been separated from each other exerts an electric force on one another with the magnitude of F. If their distance is halved, the magnitude of the force becomes F'. Calculate the value of $\frac{F'}{F}$.

Solution

Based on the information given in the problem, we have:

$$r' = 0.5r$$

The magnitude of the electric force between two electric charges (q_1 and q_2) that have been separated from each other by the distance r, can be calculated as follows.

$$F = k\frac{q_1 q_2}{r^2}$$

where, $k = \frac{1}{4\pi\varepsilon_0} = 9 \times 10^9 \frac{N.m^2}{C^2}$ is the Coulomb constant. Herein, ε_0 is the permittivity of free space which is equal to $8.85 \times 10^{-12} \frac{C^2}{N.m^2}$.

The electric force in the second case can be calculated as follows.

$$F' = k\frac{q'_1 q'_2}{r'^2}$$

$$\Rightarrow F' = k\frac{q'_1 q'_2}{(\quad)^2} = (\qquad) \times k\frac{q_1 q_2}{r^2}$$

$$\Rightarrow F' = (\qquad) \times F$$

$$\Rightarrow \frac{F'}{F} = 4$$

Problem

1.6. In the system shown in Fig. 1.3, calculate the magnitude of the total force exerted on charge Q if $q = 2\ \mu C$ and $Q = 4\ \mu C$. The Coulomb's constant is $k = \frac{1}{4\pi\varepsilon_0} = 9 \times 10^9 \frac{N.m^2}{C^2}$.

Difficulty level ○ Easy ○ Normal ● Hard

Calculation amount ○ Small ● Normal ○ Large

1) 1.7 mN
2) 4.6 mN
3) 2.9 mN
4) 2.3 mN

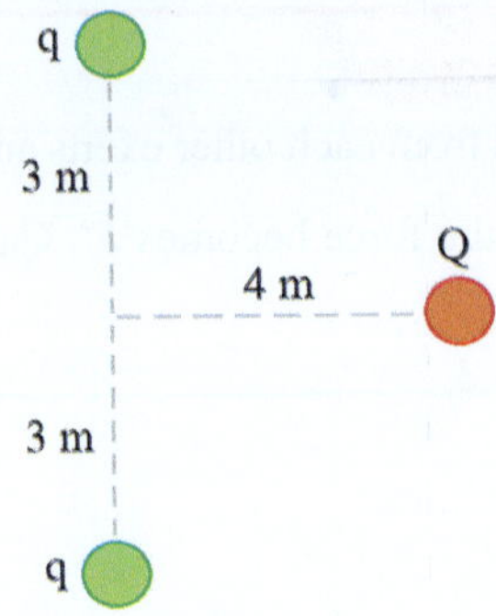

Fig. 1.3 Three charges on a plane

Partially Solved Exercise

1.7. In the system shown in Fig. 1.4, calculate the total force exerted on charge q_1 if $q_1 = q_2 = q_3 = 10\ \mu C$. The Coulomb's constant is $k = \frac{1}{4\pi\varepsilon_0} = 9 \times 10^9\ \frac{N.m^2}{C^2}$.

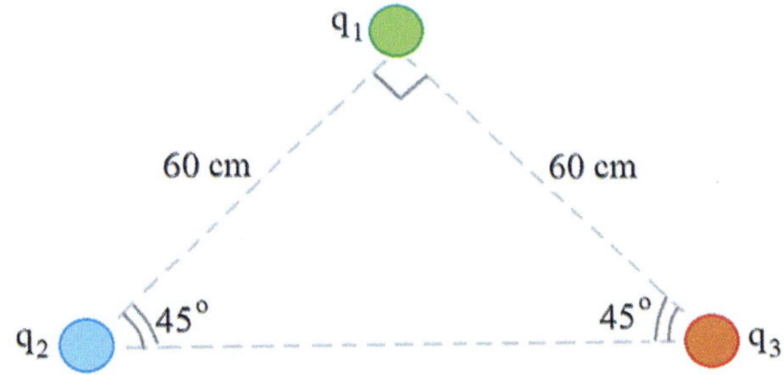

Fig. 1.4 Three charges on a plane

Solution

Based on the information given in the problem, we have:

$$q_1 = q_2 = q_3 = 10\ \mu C$$

$$k = \frac{1}{4\pi\varepsilon_0} = 9 \times 10^9\ \frac{N.m^2}{C^2}$$

Figure 1.5 shows the electric force that charges q_2 and q_3 exert on charge q_1. The magnitude of these forces can be calculated as follows.

$$F_{21} = k\frac{q_2 q_1}{r^2}$$

$$F_{31} = k\frac{q_3 q_1}{r^2}$$

$$\Rightarrow F_{21} = F_{31} = 9 \times 10^9 \times \frac{(\qquad) \times (\qquad)}{(\qquad)^2} = (\qquad)\,N$$

The magnitude of vector sum of these forces can be calculated as follows. Herein, $\alpha = 45°$, as can be noticed from Fig. 1.5.

$$F = 2F_{21} \cos \alpha$$

$$\Rightarrow F = 2 \times (\qquad) \times \cos(\qquad)° = 2 \times (\qquad) \times (\qquad)$$

$$\Rightarrow F = \frac{5\sqrt{2}}{2}$$

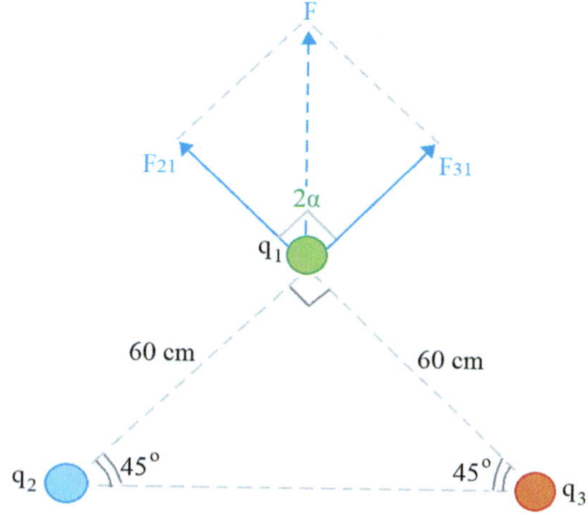

Fig. 1.5 The forces exerted on charge q_1

Notes

In this problem, the relations below have been used.

$$\cos 45° = \frac{\sqrt{2}}{2}$$

The magnitude of sum of two vectors (V_1 and V_2) that make the angle of 2α with each other can be calculated as follows.

$$V = \sqrt{V_1^2 + V_2^2 + 2V_1V_2 \cos 2\alpha}$$

If $V_1 = V_2$, we have:

$$V = 2V_1 \cos \alpha$$

1.3 Electric Field

Problem

1.8. Which one of the following choices shows wrong electric field around the positive and negative charges?

Difficulty level ● Easy ○ Normal ○ Hard 

Calculation amount ● Small ○ Normal ○ Large

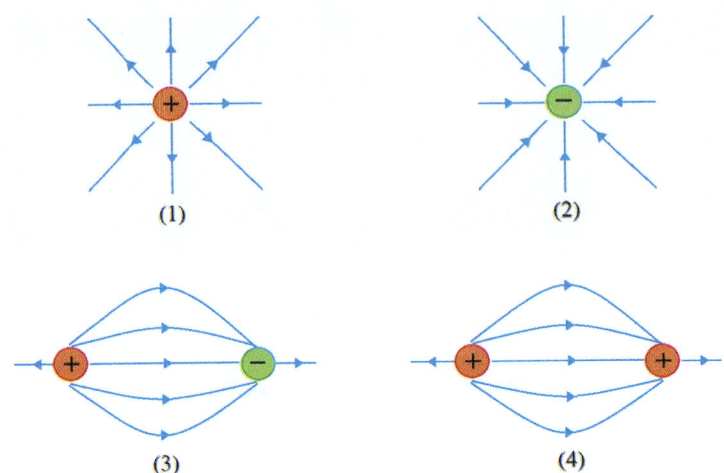

Problem

1.9. Which one of the choices below is correct about the system shown in Fig. 1.6?

Difficulty level ○ Easy ● Normal ○ Hard 

Calculation amount ○ Small ● Normal ○ Large

1) $E_A = E_B$
2) $E_A = 2E_B$
3) $E_A = 4E_B$
4) $4E_A = E_B$

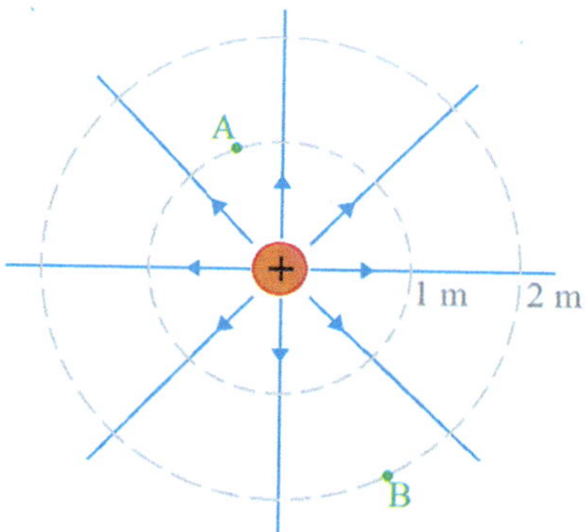

Fig. 1.6 The electric field lines around a positive charge

Exercise

1.10. Calculate the value of the following term related to the system shown in Fig. 1.7.

$$\frac{E_A}{E_B}$$

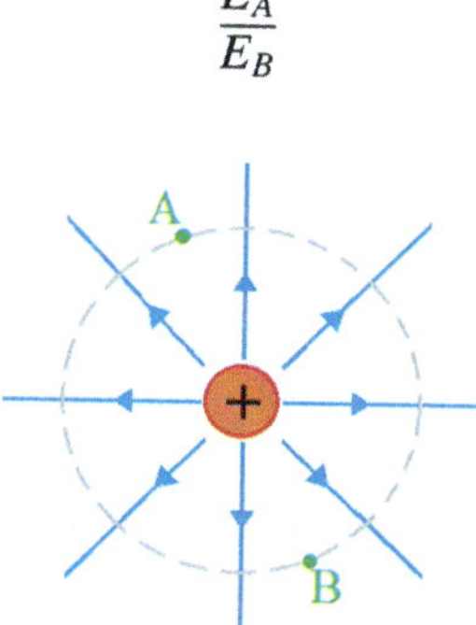

Fig. 1.7 The electric field lines around a positive charge

Final Answer

$\frac{E_A}{E_B} = 1$

Problem

1.11. The magnitude of electric filed of a point charge at a specific radial distance is E. If the magnitude of the charge and distance are tripled, the magnitude of the electric field becomes E'. Calculate the value of $\frac{E'}{E}$.

Difficulty level ○ Easy ● Normal ○ Hard
Calculation amount ○ Small ● Normal ○ Large

1) $\frac{1}{3}$
2) $\frac{1}{9}$
3) 3
4) 1

Problem

1.12. In the system shown in Fig. 1.8, calculate the magnitude of the total electric field at point p if $q = 1\ \mu C$. The Coulomb's constant is $k = \frac{1}{4\pi\varepsilon_0} = 9 \times 10^9\ \frac{N.m^2}{C^2}$.

Difficulty level ○ Easy ○ Normal ● Hard 

Calculation amount ○ Small ● Normal ○ Large

1) 576 N/C
2) 360 N/C
3) 720 N/C
4) 0 N/C

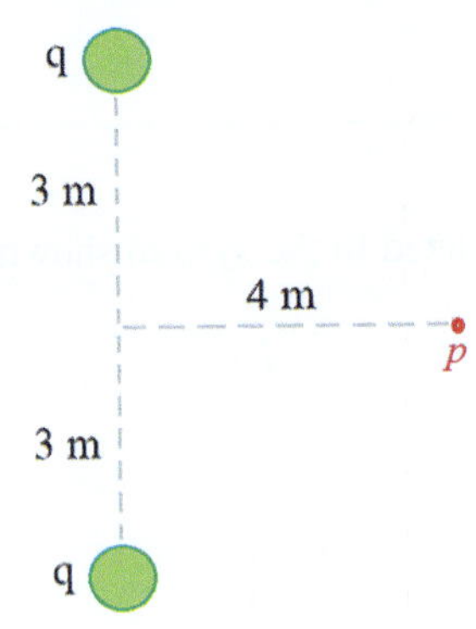

Fig. 1.8 Two charges and a point on a plane

Problem

1.13. Which one of the choices below is correct about the system shown in Fig. 1.9?

Difficulty level ○ Easy ○ Normal ● Hard

Calculation amount ● Small ○ Normal ○ Large

1) $q_1q_2 > 0, |q_1| > |q_2|$
2) $q_1q_2 < 0, |q_1| > |q_2|$
3) $q_1q_2 < 0, |q_1| < |q_2|$
4) $q_1q_2 > 0, |q_1| < |q_2|$

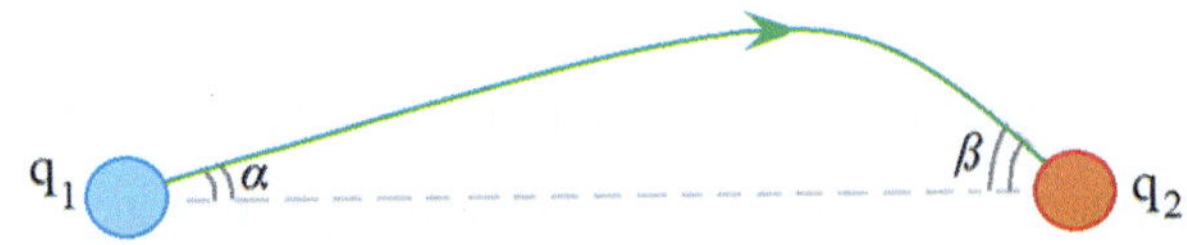

Fig. 1.9 An electric filed line around two charges

Exercise

1.14. Determine the location where the total electric field is zero around two point charges of similar type.

Final Answer

The total electric field is zero between the charges of similar type along the axis connecting them together and close to the charge with smaller magnitude.

Exercise

1.15. Determine the location where the net electric field is zero around two point charges of different types.

Final Answer

The net electric field is zero outside the charges of different types along the axis connecting them together and close to the charge with smaller magnitude.

Problem

1.16. In the system shown in Fig. 1.10, $q_1 = q$ and $q_2 = 4q$. Determine the coordinates of a point at which the total electric field is zero.

Difficulty level	○ Easy	○ Normal	● Hard
Calculation amount	○ Small	● Normal	○ Large

1) $\left(\frac{3}{4}d, 0\right)$
2) $\left(\frac{1}{4}d, 0\right)$
3) $\left(\frac{2}{3}d, 0\right)$
4) $\left(\frac{1}{3}d, 0\right)$

Fig. 1.10 Two charges of similar type in the rectangular coordinates system

Problem

1.17. In the system shown in Fig. 1.11, $q_1 = -q$ and $q_2 = 4q$. Determine the coordinates of a point at which the total electric field is zero.

Difficulty level ○ Easy ○ Normal ● Hard

Calculation amount ○ Small ● Normal ○ Large

1) $(-d, 0)$
2) $\left(\frac{1}{2}d, 0\right)$
3) $\left(-\frac{1}{3}d, 0\right)$
4) $(d, 0)$

Fig. 1.11 Two charges of different types on the rectangular coordinates system

Partially Solved Exercise

1.18. In Problem 2.16, assume that the point charges are $q_1 = q$ and $q_2 = 9q$. Determine the coordinates of a point at which the total electric field is zero.

Solution

The charges are of the same type. Thus, the total electric field is zero between them along the axis connecting them together (see Fig. 1.12).

The electric fields at point p are in opposite directions but equal.

$$E_{1p} = E_{2p}$$

$$k\frac{q_1}{x^2} = k\frac{q_2}{(d-x)^2} \Rightarrow \frac{(\quad)}{x^2} = \frac{(\quad)}{(d-x)^2}$$

$$\Rightarrow \frac{(\quad)}{x^2} = \frac{(\quad)}{(d-x)^2} \Rightarrow \frac{(\quad)}{x} = \frac{(\quad)}{d-x}$$

$$\Rightarrow$$

$$\Rightarrow (x, y) = \left(\frac{1}{4}d, 0\right)$$

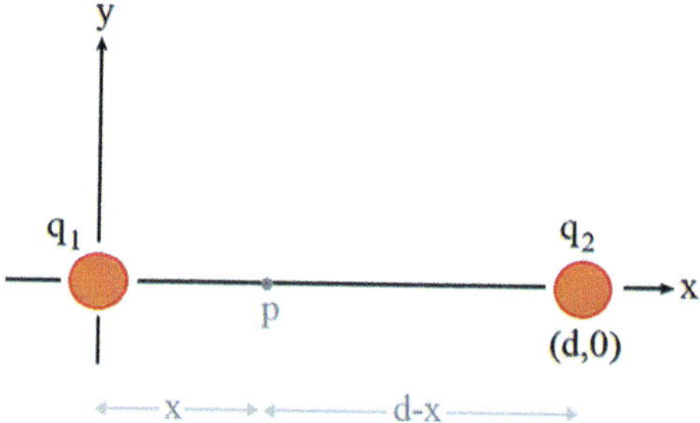

Fig. 1.12 The coordinates of a point at which the total electric field is zero

Notes

In this problem, the relations below have been used.

The magnitude of electric filed around a point charge can be calculated as follows.

$$E = k\frac{q}{r^2}$$

Herein, q is the point charge (C), r is the radial distance (m) from the charge, and $k = \frac{1}{4\pi\varepsilon_0} = 9 \times 10^9 \frac{N.m^2}{C^2}$ is the Coulomb constant. Moreover, ε_0 is the permittivity of free space which is equal to $8.85 \times 10^{-12} \frac{C^2}{N.m^2}$.

Partially Solved Exercise

1.19. In Problem 2.17, assume that the point charges are $q_1 = -q$ and $q_2 = 9q$. Determine the coordinates of a point at which the total electric field is zero.

Solution

The charges are of different types. Hence, the total electric field is zero outside them along the axis connecting them together (see Fig. 1.13).

The electric fields at point p are in opposite directions but equal.

$$E_{1p} = E_{2p}$$

$$k\frac{q_1}{x^2} = k\frac{q_2}{(d+x)^2} \Rightarrow \frac{(\quad)}{x^2} = \frac{(\quad)}{(d+x)^2}$$

$$\Rightarrow \frac{(\quad)}{x^2} = \frac{(\quad)}{(d+x)^2} \Rightarrow \frac{(\quad)}{x} = \frac{(\quad)}{d+x}$$

$$\Rightarrow$$

$$\Rightarrow (x,y) = \left(-\frac{d}{2}, 0\right)$$

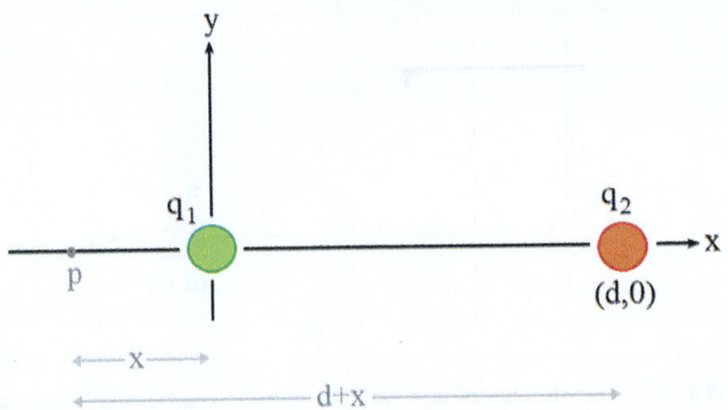

Fig. 1.13 The coordinates of a point where the total electric field is zero

Notes

In this problem, the relations below have been used.

The magnitude of electric filed around a point charge (q) can be calculated as follows.

$$E = k\frac{q}{r^2}$$

Herein, r is the radial distance from the point charge and $k = \frac{1}{4\pi\varepsilon_0} = 9 \times 10^9 \frac{N.m^2}{C^2}$ is the Coulomb constant. Moreover, ε_0 is the permittivity of free space which is equal to $8.85 \times 10^{-12} \frac{C^2}{N.m^2}$.

Problem

1.20. Which one of the following choices shows the correct direction for the electric forces on the positive and negative charges in the uniform electric field (Fig. 1.14)?

Difficulty level ○ Easy ○ Normal ● Hard 

Calculation amount ● Small ○ Normal ○ Large

1) Only F_1
2) Only F_3
3) Both F_2 and F_3
4) Both F_1 and F_3

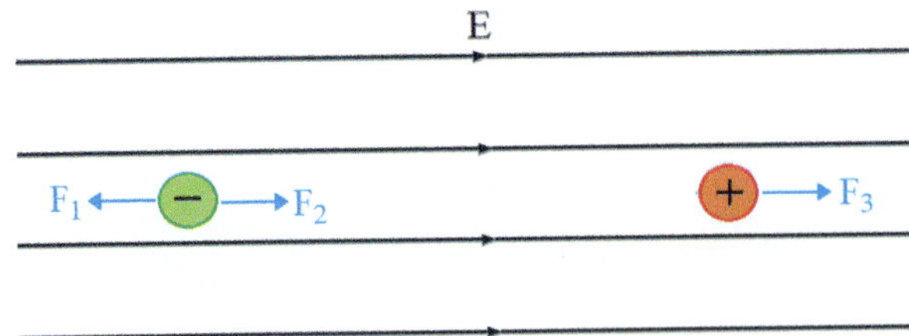

Fig. 1.14 The electric forces experienced by the positive and negative charges in the uniform electric field

Problem

1.21. A weightless particle with the charge of $q = -6 \times 10^{-9}\ C$ senses a downward uniform electric force with the magnitude of $F = 3 \times 10^{-6}\ N$. Determine the magnitude and direction of the electric field.

Difficulty level ○ Easy ● Normal ○ Hard
Calculation amount ● Small ○ Normal ○ Large

1) 500 N/C, upward
2) $2 \times 10^{-3}\ N/C$, upward
3) 500 N/C, downward
4) $2 \times 10^{-3}\ N/C$, downward

Exercise

1.22. A weightless particle with the charge of $q = 10^{-9}\ C$ experiences a downward uniform electric force with the magnitude of $F = 10^{-6}$. Determine the magnitude and direction of the electric field.

Final Answer

1000 N/C, downward

Problem

1.23. A stationary electron is freed in a uniform electric field with the magnitude of $2 \times 10^{4}\ N/C$. Calculate the acceleration of the electron. The mass and charge of an electron are about $9.11 \times 10^{-31}\ kg$ and $-1.6 \times 10^{-19}\ C$, respectively. Herein, ignore the gravitational force since the subatomic particles such as electron and proton are too small to experience it.

Difficulty level ○ Easy ● Normal ○ Hard
Calculation amount ● Small ○ Normal ○ Large

1) $351 \times 10^{12}\ m/s^2$
2) $351 \times 10^{13}\ m/s^2$
3) $351 \times 10^{14}\ m/s^2$
4) $351 \times 10^{15}\ m/s^2$

Problem

1.24. In the system shown in Fig. 1.15, a positively charged particle with the mass of m is suspended in the air and in an uniform electric field. Determine the mass of the particle that keeps it in this position.

1) $\dfrac{Vq}{dg}$
2) $\dfrac{Vg}{dq}$
3) $\dfrac{dq}{Vg}$
4) $\dfrac{Vqg}{d}$

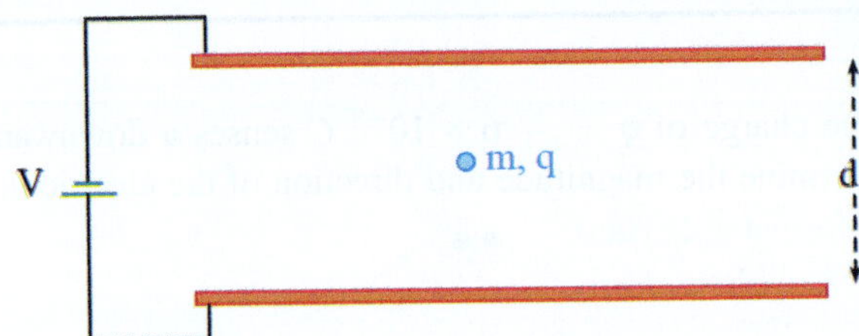

Fig. 1.15 A positively charged particle suspended in the air and in a uniform electric field

Problem

1.25. Calculate the electric field of a very large surface with the surface charge density of $\sigma = 10.6\ \mu C/m^2$, shown in Fig. 1.16. The permittivity of free space is $\varepsilon_0 = 8.85 \times 10^{-12} \frac{C^2}{N.m^2}$.

Difficulty level ● Easy ○ Normal ○ Hard

Calculation amount ● Small ○ Normal ○ Large

1) $6 \times 10^6\ N/C$
2) $6 \times 10^5\ N/C$
3) $12 \times 10^6\ N/C$
4) $12 \times 10^5\ N/C$

Fig. 1.16 A very large surface with the surface charge density

1.4 Electric Potential

Problem

1.26. The electric potential between two points is about 3000 V. How much electric charge can be moved between these two points with the 0.6 mJ energy?

Difficulty level ● Easy ○ Normal ○ Hard

Calculation amount ● Small ○ Normal ○ Large

1) 0.1 μC
2) 0.2 μC
3) 0.45 μC
4) 4 μC

Problem

1.27. In Fig. 1.17, $E = 3000$ *N/C* and the distance between the two points is 2 *cm*. Calculate the value of $V_{12} = V_1 - V_2$.

Difficulty level ○ Easy ● Normal ○ Hard

Calculation amount ● Small ○ Normal ○ Large

1) -6000 *V*
2) 6000 *V*
3) -60 *V*
4) 60 *V*

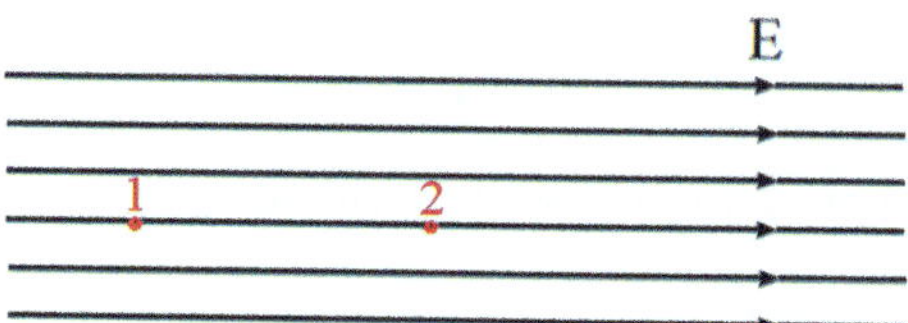

Fig. 1.17 A uniform electric field

Problem

1.28. Which one of the choices below is correct about the system shown in Fig. 1.18?

Difficulty level ○ Easy ● Normal ○ Hard

Calculation amount ○ Small ● Normal ○ Large

1) $V_A = V_B$
2) $V_A = 2V_B$
3) $V_A = 4V_B$
4) $2V_A = V_B$

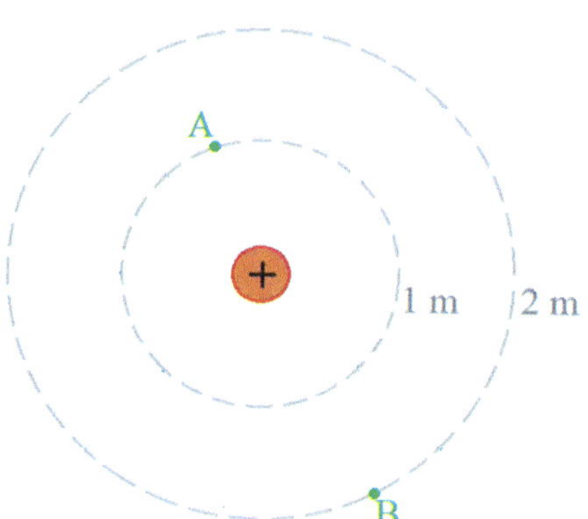

Fig. 1.18 The equipotential lines around a positive charge

Exercise

1.29. Calculate the value of $\frac{V_A}{V_B}$ related to the system shown in Fig. 1.19.

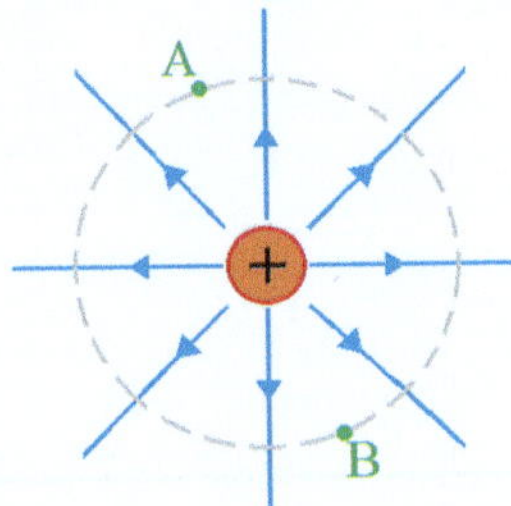

Fig. 1.19 An equipotential line around a positive charge

Final Answer

$\frac{V_A}{V_B} = 1$

Problem

1.30. The electric potential of a point charge at a specific radial distance is V. If the magnitude of the charge and distance are tripled, the electric potential becomes V'. Calculate the value of $\frac{V'}{V}$.

Difficulty level ○ Easy ● Normal ○ Hard 

Calculation amount ○ Small ● Normal ○ Large

1) $\frac{1}{3}$
2) 1
3) 3
4) $\frac{1}{9}$

Problem

1.31. In the system shown in Fig. 1.20, calculate the total electric potential at point p if $q = 1\ \mu C$. The Coulomb's constant is $k = \frac{1}{4\pi\varepsilon_0} = 9 \times 10^9\ \frac{N.m^2}{C^2}$.

Difficulty level ○ Easy ● Normal ○ Hard 

Calculation amount ○ Small ● Normal ○ Large

1) 0 V
2) 360 V
3) 720 V
4) 3600 V

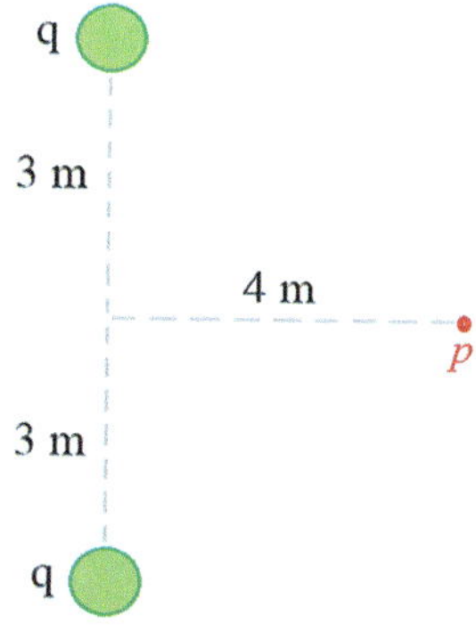

Fig. 1.20 Two charges and a point on a plane

Exercise

1.32. In the system shown in Fig. 1.21, calculate the total electric potential at point p if $q_1 = 1\ \mu C$ and $q_2 = -1\ \mu C$. The Coulomb's constant is $k = \frac{1}{4\pi\varepsilon_0} = 9 \times 10^9\ \frac{N.m^2}{C^2}$.

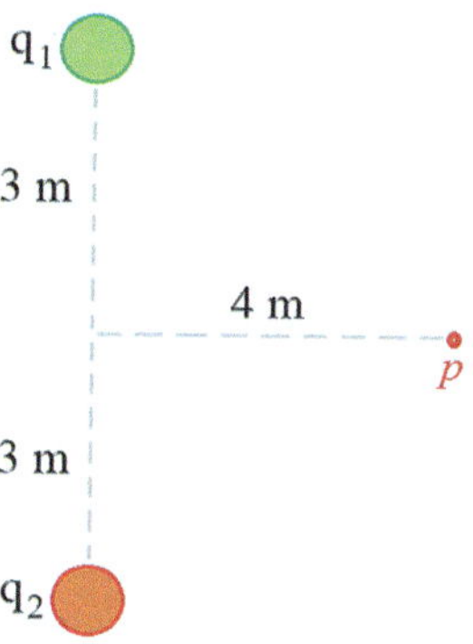

Fig. 1.21 Two charges and a point on a plane

Final Answer

$V = 0\ V$

Problem

1.33. Calculate total electric potential at the center of the square shown in Fig. 1.22. Herein, we have:

$$a = \sqrt{2}\ m$$

$$q_1 = 12\ nC$$

$$q_2 = -24\ nC$$

$$q_3 = 31\ nC$$

$$q_4 = 17\ nC$$

The Coulomb's constant is $k = \frac{1}{4\pi\varepsilon_0} = 9 \times 10^9\ \frac{N.m^2}{C^2}$.

Difficulty level ○ Easy ○ Normal ● Hard 

Calculation amount ○ Small ● Normal ○ Large

1) $162 \times 10^9\ V$
2) $9 \times 10^9\ V$
3) $-9 \times 10^9\ V$
4) $0\ V$

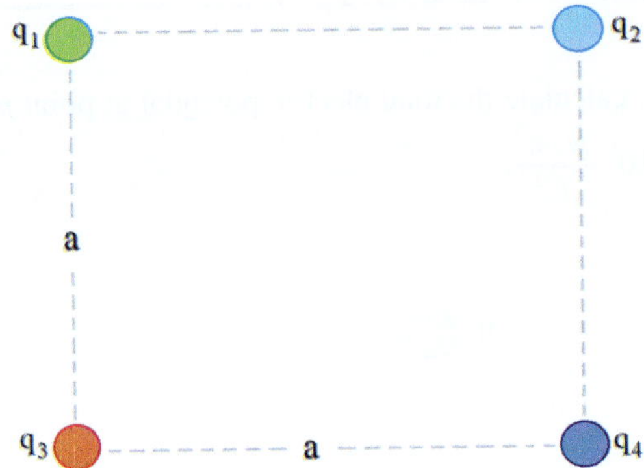

Fig. 1.22 Four charges on the corners (vertices) of a square

Exercise

1.34. Determine the location where the net electric potential is zero around two point charges of similar type.

Final Answer

There is no place where the net electric potential is zero.

Exercise

1.35. Determine the location where the electric potential is zero around two point charges of different types.

Final Answer

The bisector plane of the line segment connecting the charges together is the area where the electric potential is zero (See Fig. 1.23).

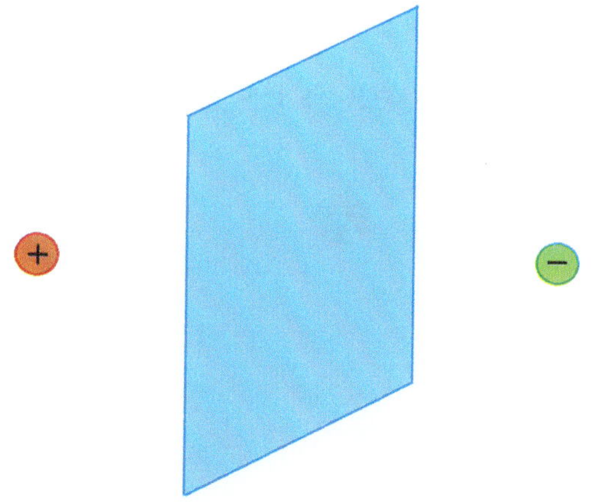

Fig. 1.23 The location where the electric potential is zero around two point charges of different types

Problem

1.36. Which one of the following choices shows the correct equipotential surfaces around the positive and negative point charges?

Difficulty level ● Easy ○ Normal ○ Hard
Calculation amount ● Small ○ Normal ○ Large

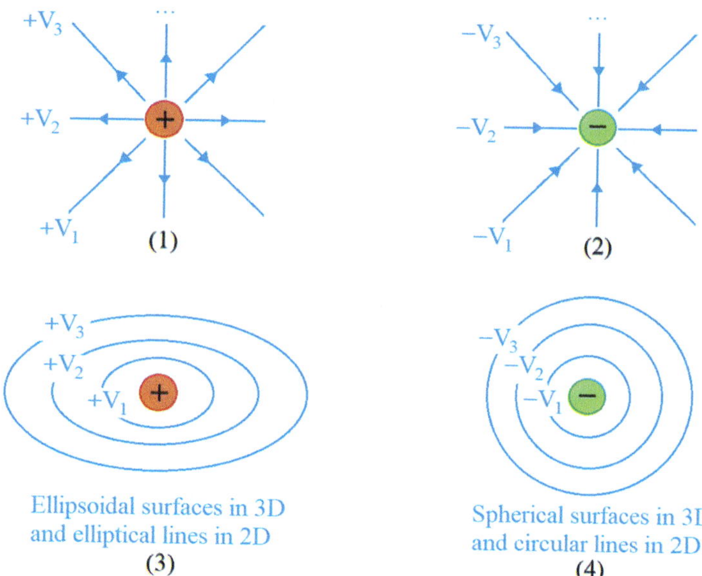

Problem

1.37. Calculate the electric potential difference between the points A and B in the system shown in Fig. 1.24. The Coulomb's constant is $k = \frac{1}{4\pi\varepsilon_0} = 9 \times 10^9 \frac{N.m^2}{C^2}$.

Difficulty level ○ Easy ○ Normal ● Hard 

Calculation amount ○ Small ● Normal ○ Large

1) 0
2) $\frac{3q}{16\pi\varepsilon_0 d}$
3) $\frac{3q}{8\pi\varepsilon_0 d}$
4) $\frac{3q}{4\pi\varepsilon_0 d}$

Fig. 1.24 Two point charges and two points on the corners (vertices) of a rectangle

1.5 Electric Potential Energy

Problem

1.38. Calculate the electric potential energy of two 1 μC charges separated by two meters. The Coulomb's constant is $k = \frac{1}{4\pi\varepsilon_0} = 9 \times 10^9 \frac{N.m^2}{C^2}$.

Difficulty level ○ Easy ● Normal ○ Hard 

Calculation amount ● Small ○ Normal ○ Large

1) 2.25 J
2) 2.25 mJ
3) 4.5 J
4) 4.5 mJ

Problem

1.39. Calculate the electric potential energy of the system shown in Fig. 1.25. Herein, we have:

$$d = 12\ cm$$

$$q_1 = 15\ \mu C$$

$$q_2 = -60\ \mu C$$

$$q_3 = 30\ \mu C$$

The Coulomb's constant is $k = \frac{1}{4\pi\varepsilon_0} = 9 \times 10^9\ \frac{N.m^2}{C^2}$.

Difficulty level ○ Easy ○ Normal ● Hard 

Calculation amount ○ Small ● Normal ○ Large

1) $-168.75\ J$
2) $168.75\ J$
3) $168.75\ mJ$
4) $-168.75\ mJ$

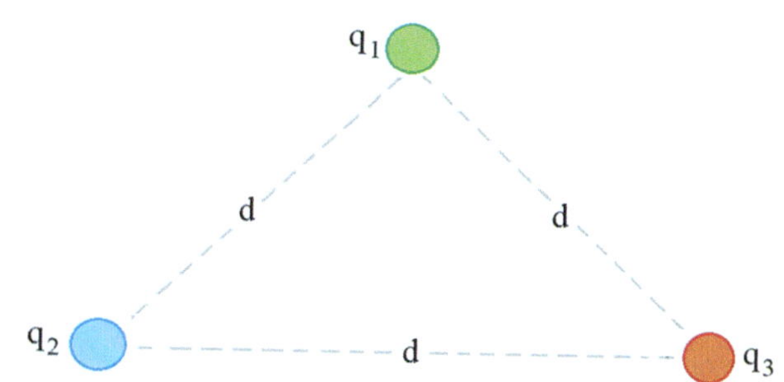

Fig. 1.25 Three charges on the corners (vertices) of an equilateral triangle

Partially Solved Exercise

1.40. Three charges are placed at the corners (vertices) of an equilateral triangle with one meter side length shown in Fig. 1.26. Determine the amount of work needed to move them to the other equilateral triangle with 25 cm side length shown in Fig. 1.26. Herein, $q_1 = 1\ \mu C$, $q_2 = 2\ \mu C$ and $q_3 = 3\ \mu C$. The Coulomb's constant is $k = \frac{1}{4\pi\varepsilon_0} = 9 \times 10^9\ \frac{N.m^2}{C^2}$.

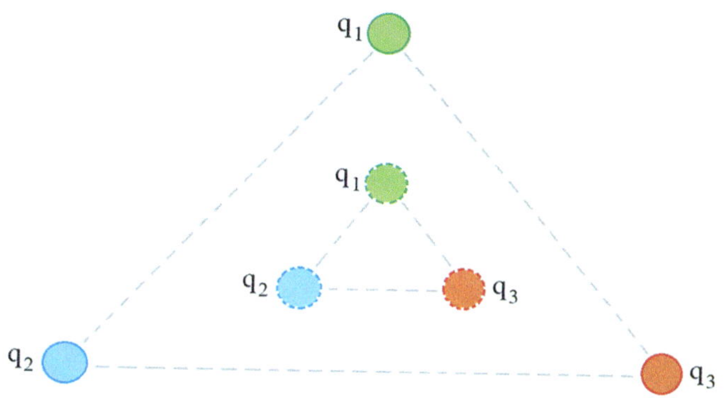

Fig. 1.26 Three charges move from the corners (vertices) of an equilateral triangle to the vertices of the smaller one

Solution

Based on the information given in the problem, we have:

$$q_1 = 1\ \mu C$$

$$q_2 = 2\ \mu C$$

$$q_3 = 3\ \mu C$$

$$r_{12} = r_{23} = r_{13} = 1\ m$$

$$r'_{12} = r'_{23} = r'_{13} = 0.25\ m$$

$$k = \frac{1}{4\pi\varepsilon_0} = 9\times 10^9\ \frac{N.m^2}{C^2}$$

The amount of work needed to move the charges can be calculated as follows.

$$W = \Delta U = U' - U$$

Moreover, the electric potential energy of a system of three charges can be calculated as follows.

$$U = U_{12} + U_{13} + U_{23}$$

$$\Rightarrow U = k\frac{q_1 q_2}{r_{12}} + k\frac{q_1 q_3}{r_{13}} + k\frac{q_2 q_3}{r_{23}}$$

For the initial state of the system, we have:

$$U = \frac{1}{4\pi\varepsilon_0}\left(\frac{(\qquad)\times(\qquad)}{(\qquad)} + \frac{(\qquad)\times(\qquad)}{(\qquad)} + \frac{(\qquad)\times(\qquad)}{(\qquad)}\right)$$

$$\Rightarrow U = \frac{(\qquad)}{4\pi\varepsilon_0}\ J$$

For the final state of the system, we have:

$$U' = \frac{1}{4\pi\varepsilon_0}\left(\frac{(\qquad)\times(\qquad)}{(\qquad)} + \frac{(\qquad)\times(\qquad)}{(\qquad)} + \frac{(\qquad)\times(\qquad)}{(\qquad)}\right)$$

$$\Rightarrow U' = \frac{(\qquad)}{\pi\varepsilon_0}\ J$$

The amount of work needed to move the charges:

$$W = U' - U = \frac{(\qquad)}{\pi\varepsilon_0} - \frac{(\qquad)}{4\pi\varepsilon_0}$$

$$\Rightarrow W = \frac{33\times 10^{-12}}{4\pi\varepsilon_0}\ J$$

Problem

1.41. In the system shown in Fig. 1.27, calculate the amount of work needed to move charge q' from point A to point B. The Coulomb's constant is $k = \frac{1}{4\pi\varepsilon_0} = 9 \times 10^9 \frac{N.m^2}{C^2}$.

Difficulty level ○ Easy ○ Normal ● Hard

Calculation amount ○ Small ● Normal ○ Large

1) 0
2) $\frac{kq'q}{a}\left(1 + \frac{1}{\sqrt{2}}\right)$
3) $\frac{kq'q}{a}\left(1 - \frac{1}{\sqrt{2}}\right)$
4) $\frac{2kq'q}{a}$

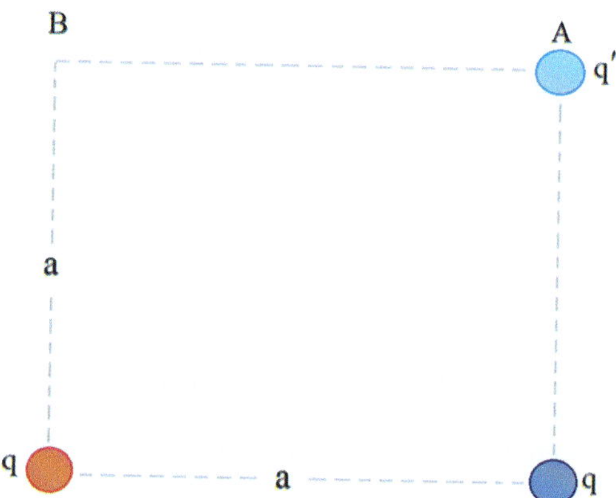

Fig. 1.27 Three point charges and one point on the corners (vertices) of a square

1.6 Electric Energy and Electric Energy Density

Problem

1.42. A uniform electric field with the strength of 10^3 N/C exists inside a cuboid capacitor. Calculate the electric energy density inside the capacitor located in a free space. Herein, assume that $\varepsilon_0 = 8.85 \times 10^{-12} \frac{C^2}{N.m^2}$.

Difficulty level ● Easy ○ Normal ○ Hard

Calculation amount ● Small ○ Normal ○ Large

1) $\omega = 0\ J/m^3$
2) $\omega = 8.85\ \mu J/m^3$
3) $\omega = 2.21\ \mu J/m^3$
4) $\omega = 4.42\ \mu J/m^3$

Problem

1.43. In Problem 1.42, calculate the total electric energy stored inside the capacitor if its dimensions are $2 \times 1 \times 0.2\ cm$.

Difficulty level ● Easy ○ Normal ○ Hard

Calculation amount ● Small ○ Normal ○ Large

1) $W = 17.68\ pJ$
2) $W = 0.1768\ pJ$
3) $W = 176.8\ pJ$
4) $W = 1.768\ pJ$

Exercise

1.44. In an environment, a uniform electric field with the strength of $\sqrt{2}\,N/C$ is available. Calculate the electric energy density and total energy stored in a free space with the dimensions of $1 \times 1 \times 1\ m$

Final Answer

$\omega = \varepsilon_0\ J/m^3$, $W = \varepsilon_0\ J$

References

1. Rahmani-Andebili, M., General Physics I - Practice Problems, Methods, and Solutions, Springer Nature, 2025.
2. Rahmani-Andebili, M., Calculus III – Practice Problems, Methods, and Solutions, Springer Nature, 2023.
3. Rahmani-Andebili, M., Calculus II – Practice Problems, Methods, and Solutions, Springer Nature, 2023.
4. Rahmani-Andebili, M., Calculus I (2nd Ed.) – Practice Problems, Methods, and Solutions, Springer Nature, 2023.
5. Rahmani-Andebili, M., Precalculus (2nd Ed.) – Practice Problems, Methods, and Solutions, Springer Nature, 2024.

Electrostatics: Part B

2

Abstract

In this chapter, the problems of the first chapter are fully solved, in detail, step-by-step, and with different methods.

2.1 Electric Charge

2.1. Based on the information given in the problem, we have [1–5]:

$$q = 4\ C$$

$$e = 1.6 \times 10^{-19}\ C$$

As we know:

$$q = ne$$

Therefore:

$$n = \frac{q}{e} = \frac{4}{1.6 \times 10^{-19}}$$

$$\Rightarrow n = 2500 \times 10^{16}$$

Choice (3) is the answer.

2.2 Electric Force

2.2. As we know, the electric force between two similar charges (positive-positive charges or negative-negative charges) is repulsive and between two different charges (positive-negative charges or negative-positive charges) is attractive. Thus, the one shown in Choice (3) is wrong since it must be attractive as is shown in Fig. 2.1.

Choice (3) is the answer.

M. Rahmani-Andebili, *General Physics II*, https://doi.org/10.1007/978-3-031-92866-6_2

Fig. 2.1 The correct electric force exerted on the test charge

---

2.3. Newton's third law states that for every action (force) in nature there is an equal and opposite reaction. Hence:

$$F_{21} = -F_{12}, \quad |F_{21}| = |F_{12}|$$

Choice (4) is the answer (Fig. 2.2).

Fig. 2.2 Two charges with unknown types. Herein, $|q_1| > |q_2|$

2.4. Based on the information given in the problem, we have:

$$r' = 2r$$

$$q'_1 = 2q_1$$

$$q'_2 = 2q_2$$

The magnitude of the electric force between two electric charges (q_1 and q_2) that have been separated from each other by the distance r, can be calculated as follows.

$$F = k\frac{q_1 q_2}{r^2}$$

where, $k = \frac{1}{4\pi\varepsilon_0} = 9 \times 10^9 \frac{N.m^2}{C^2}$ is the Coulomb constant. Herein, ε_0 is the permittivity of free space which is equal to $8.85 \times 10^{-12} \frac{C^2}{N.m^2}$.

The electric force in the second case can be calculated as follows.

$$F' = k\frac{q'_1 q'_2}{r'^2}$$

$$\Rightarrow F' = k\frac{(2q_1)(2q_2)}{(2r)^2} = k\frac{q_1 q_2}{r^2}$$

$$\Rightarrow F' = F$$

$$\Rightarrow \frac{F'}{F} = 1$$

Choice (4) is the answer.

2.6. Based on the information given in the problem, we have:

$$q = 2\ \mu C$$

$$Q = 4\ \mu C$$

$$k = \frac{1}{4\pi\varepsilon_0} = 9 \times 10^9\ \frac{N.m^2}{C^2}$$

Figure 2.3 shows the electric force that each of the charges q exerts on charge Q. The magnitude of each of these forces can be calculated as follows.

$$F_1 = F_2 = k\frac{qQ}{r^2}$$

$$\Rightarrow F_1 = F_2 = 9 \times 10^9 \times \frac{2 \times 10^{-6} \times 4 \times 10^{-6}}{3^2 + 4^2}$$

$$\Rightarrow F_1 = F_2 = 2.88 \times 10^{-3}\ N$$

The magnitude of vector sum of these forces can be calculated as follows.

$$F = 2F_1 \cos \alpha$$

$$\Rightarrow F = 2 \times 2.88 \times 10^{-3} \times \frac{4}{\sqrt{3^2 + 4^2}}$$

$$\Rightarrow F = 4.6 \times 10^{-3}\ N = 4.6\ mN$$

Choice (2) is the answer.

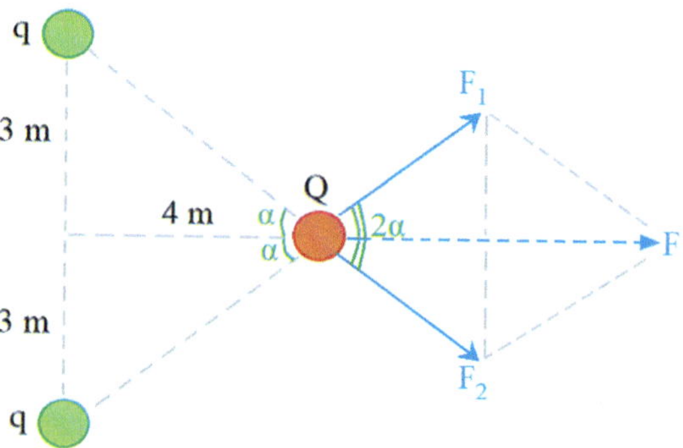

Fig. 2.3 The forces exerted on charge Q

Notes

In this problem, the relations below have been used.

$$\cos(\text{angle}) = \frac{\text{side adjacent}}{\text{hypotenuse}}$$

The magnitude of vector sum of two force vectors (F_1 and F_2) that make the angle of 2α with each other can be calculated as follows.

$$F = \sqrt{F_1^2 + F_2^2 + 2F_1F_2\cos 2\alpha}$$

If $F_1 = F_2$, we have:

$$F = 2F_1 \cos\alpha$$

2.3 Electric Field

2.8. As we know, the electric field around a positive point charge is radial and outward. Moreover, the electric field around a negative point charge is radial and inward. In addition, the electric field around two different point charges (e.g. positive-negative charges) is like the one shown in Choice (3). As can be seen, it starts from the positive charge and ends at negative charge.

However, the electric filed shown in Choice (4) is wrong. The correct one is illustrated in Fig. 2.4. As can be seen, the electric field starts from the positive charges and ends at infinity. Choice (4) is the answer.

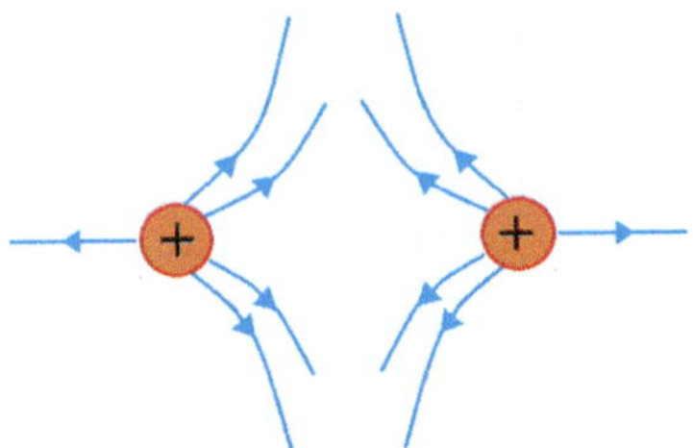

Fig. 2.4 The correct electric field around two adjacent positive charges

2.9. As we know, the magnitude of electric filed around a point charge (q) can be calculated as follows.

$$E = k\frac{q}{r^2}$$

Herein, r is the radial distance from the point charge and $k = \frac{1}{4\pi\varepsilon_0} = 9 \times 10^9 \frac{N.m^2}{C^2}$ is the Coulomb constant. Moreover, ε_0 is the permittivity of free space which is equal to $8.85 \times 10^{-12} \frac{C^2}{N.m^2}$.

Hence:

$$E_A = k\frac{q}{1^2} = kq$$

$$E_B = k\frac{q}{2^2} = \frac{1}{4}kq$$

Therefore:

$$E_A = 4E_B$$

Choice (3) is the answer (Fig. 2.5).

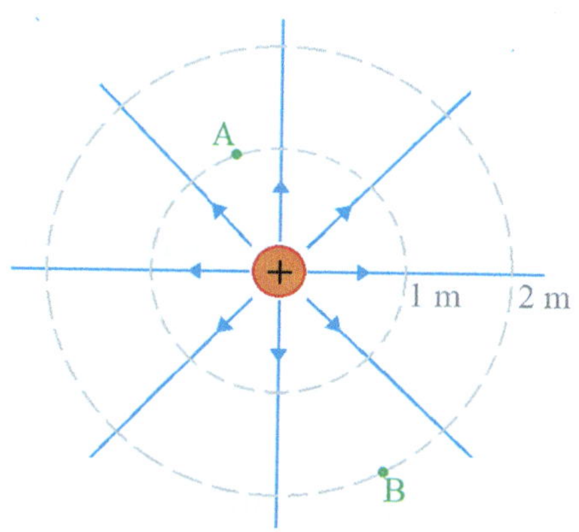

Fig. 2.5 The electric field lines around a positive charge

2.11. Based on the information given in the problem, we have:

$$r' = 3r$$

$$q' = 3q$$

As we know, the magnitude of electric filed around a point charge (q) at the radial distance r can be calculated as follows.

$$E = k\frac{q}{r^2}$$

where, $k = \frac{1}{4\pi\varepsilon_0} = 9 \times 10^9\ \frac{N.m^2}{C^2}$ is the Coulomb constant. Herein, ε_0 is the permittivity of free space which is equal to $8.85 \times 10^{-12}\frac{C^2}{N.m^2}$.

The magnitude of electric filed in the second case can be calculated as follows.

$$E' = k\frac{q'}{r'^2}$$

$$\Rightarrow E' = k\frac{3q}{(3r)^2} = \frac{1}{3}k\frac{q}{r^2}$$

$$\Rightarrow E' = \frac{1}{3}E$$

$$\Rightarrow \frac{E'}{E} = \frac{1}{3}$$

Choice (1) is the answer.

2.12. Based on the information given in the problem, we have:

$$q = 1\ \mu C$$

$$k = \frac{1}{4\pi\varepsilon_0} = 9 \times 10^9\ \frac{N.m^2}{C^2}$$

Figure 2.6 shows the electric field that each of the charges generate at point p. The magnitude of each of these fields can be calculated as follows.

$$E_1 = E_2 = k\frac{q}{r^2}$$

$$\Rightarrow E_1 = E_2 = 9 \times 10^9 \times \frac{10^{-6}}{3^2 + 4^2}$$

$$\Rightarrow E_1 = E_2 = 360\ N/C$$

The magnitude of vector sum of these fields can be calculated as follows.

$$E = 2E_1 \cos\alpha$$

$$\Rightarrow E = 2 \times 360 \times \frac{4}{\sqrt{3^2 + 4^2}}$$

$$\Rightarrow E = 576\ N/C$$

Choice (1) is the answer.

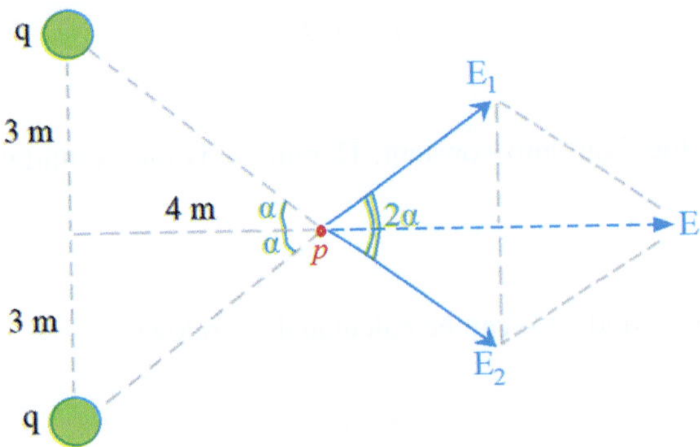

Fig. 2.6 The electric fields at point p

Notes

In this problem, the relations below have been used.

$$\cos(\text{angle}) = \frac{\text{side adjacent}}{\text{hypotenuse}}$$

The magnitude of sum of two vectors (V_1 and V_2) that make the angle of 2α with each other can be calculated as follows.

$$V = \sqrt{V_1^2 + V_2^2 + 2V_1V_2\cos 2\alpha}$$

If $V_1 = V_2$, we have:

$$V = 2V_1 \cos\alpha$$

2.13. Figure 2.7 shows the system. Based on the direction of the electric field, q_1 is a positive charge and q_2 is a negative charge. Therefore, the type of charges is different. In other words:

$$q_1 q_2 < 0$$

Moreover, since $\alpha < \beta$, the electric field around q_1 is stronger than the one around q_2. Hence:

$$|q_1| > |q_2|$$

Choice (2) is the answer.

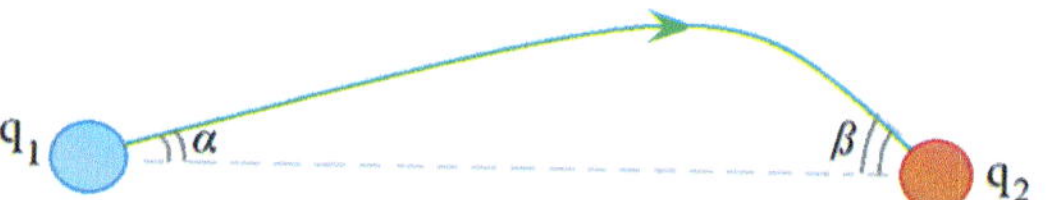

Fig. 2.7 An electric filed line around two charges

2.16. Based on the information given in the problem, we have:

$$q_1 = q$$

$$q_2 = 4q$$

The charges are of the same type. Thus, the total electric field is zero between them along the axis connecting them together (see Fig. 2.8).

The electric fields at point p are in opposite directions but equal. Thus:

$$E_{1p} = E_{2p}$$

$$k\frac{q_1}{x^2} = k\frac{q_2}{(d-x)^2} \Rightarrow \frac{q}{x^2} = \frac{4q}{(d-x)^2}$$

$$\Rightarrow \frac{1}{x^2} = \frac{4}{(d-x)^2} \Rightarrow \frac{1}{x} = \frac{2}{d-x}$$

$$\Rightarrow 2x = d - x \Rightarrow x = \frac{d}{3}$$

$$\Rightarrow (x, y) = \left(\frac{1}{3}d, 0\right)$$

Choice (4) is the answer.

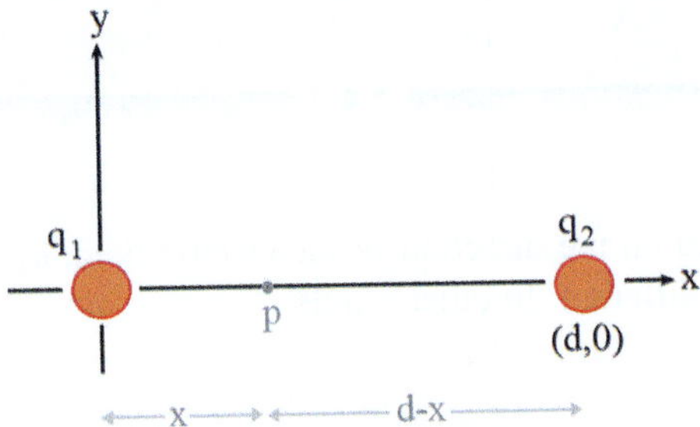

Fig. 2.8 The coordinates of a point where the total electric field is zero

Notes

In this problem, the relations below have been used.

The magnitude of electric filed around a point charge can be calculated as follows.

$$E = k\frac{q}{r^2}$$

Herein, q is the point charge (C), r is the radial distance (m) from the charge, and $k = \frac{1}{4\pi\varepsilon_0} = 9 \times 10^9 \frac{N.m^2}{C^2}$ is the Coulomb constant. Moreover, ε_0 is the permittivity of free space which is equal to $8.85 \times 10^{-12} \frac{C^2}{N.m^2}$.

2.17. Based on the information given in the problem, we have:

$$q_1 = -q$$

$$q_2 = 4q$$

The charges are of different types. Hence, the total electric field is zero outside them along the axis connecting them together (see Fig. 2.9).

The electric fields at point p are in opposite directions but equal. Hence:

$$E_{1p} = E_{2p}$$

$$k\frac{q_1}{x^2} = k\frac{q_2}{(d+x)^2} \Rightarrow \frac{q}{x^2} = \frac{4q}{(d+x)^2}$$

$$\Rightarrow \frac{1}{x^2} = \frac{4}{(d+x)^2} \Rightarrow \frac{1}{x} = \frac{2}{d+x}$$

$$\Rightarrow 2x = d + x \Rightarrow x = d$$

$$\Rightarrow (x, y) = (-d, 0)$$

Choice (1) is the answer.

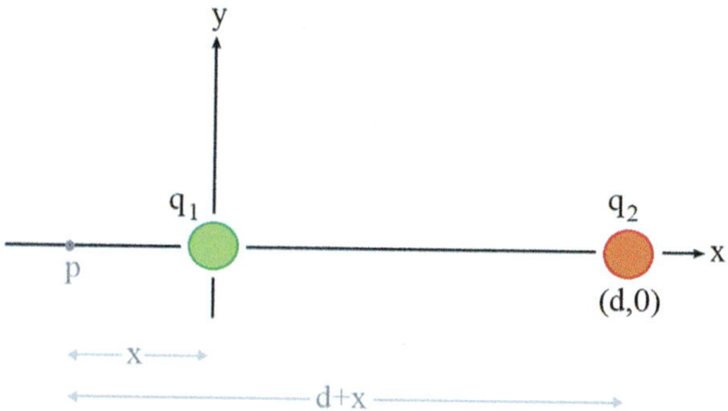

Fig. 2.9 The coordinates of a point where the total electric field is zero

Notes

In this problem, the relations below have been used.

The magnitude of electric filed around a point charge (q) can be calculated as follows.

$$E = k\frac{q}{r^2}$$

Herein, r is the radial distance from the point charge and $k = \frac{1}{4\pi\varepsilon_0} = 9 \times 10^9\ \frac{N.m^2}{C^2}$ is the Coulomb constant. Moreover, ε_0 is the permittivity of free space which is equal to $8.85 \times 10^{-12}\ \frac{C^2}{N.m^2}$.

2.20. The direction of electric force on a positive charge is in the same direction of the electric field since $\vec{F} = (+q)\vec{E}$. Thus, F_3 is correct.

Moreover, the direction of electric force on a negative charge is in the opposite direction of the electric field since $\vec{F} = (-q)\vec{E}$. Hence, F_1 is correct.

Choice (4) is the answer (Fig. 2.10).

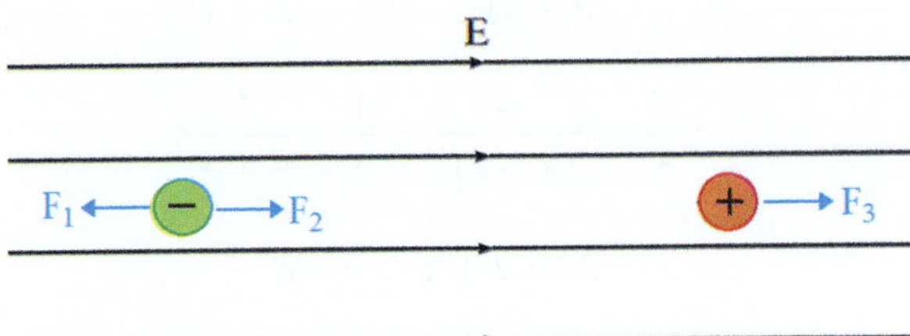

Fig. 2.10 The electric forces experienced by the positive and negative charges in the uniform electric field

---

2.21. Based on the information given in the problem, we have:

$$q = -6 \times 10^{-9}\ C$$

$$F = 3 \times 10^{-6}\ N$$

The magnitude of the uniform electric field can be calculated as follows.

$$F = E|q| \Rightarrow E = \frac{F}{|q|}$$

$$\Rightarrow E = \frac{3 \times 10^{-6}}{6 \times 10^{-9}} \Rightarrow E = 500\ N/C$$

On the other hand, as we know, the direction of electric force on a negative charge is in the opposite direction of the electric field. Therefore, the electric field is upward.

Choice (1) is the answer.

2.23. Based on the information given in the problem, we have:

$$E = 2 \times 10^{4}\ N/C$$

$$m = 9.11 \times 10^{-31}\ kg$$

$$q = -1.6 \times 10^{-19}\ C$$

The only force that the electron senses is the electric force since the gravitational force is too small for an electron.

Based on Newton's second law, we have:

$$\sum F = ma$$

$$\Rightarrow E|q| = ma$$

$$\Rightarrow a = \frac{E|q|}{m} = \frac{1.6 \times 10^{-19} \times 2 \times 10^4}{9.11 \times 10^{-31}}$$

$$\Rightarrow a = 351 \times 10^{13}\ \frac{m}{s^2}$$

Choice (2) is the answer.

2.24. As can be seen in Fig. 2.11, the uniform electric field is upward due to the polarity of voltage source. The relation between a uniform electric field and voltage is as follows.

$$E = \frac{V}{d}$$

Moreover, two forces, that is, upward electric force and downward gravitational force are exerted on the particle. The particle is at a stationary position. Hence, based on Newton's first law, we have:

$$\sum F = 0$$

$$\Rightarrow F_g - F_E = 0 \Rightarrow F_g = F_E$$

$$\Rightarrow mg = Eq \Rightarrow mg = \frac{V}{d}q$$

$$\Rightarrow m = \frac{Vq}{dg}$$

Choice (1) is the answer.

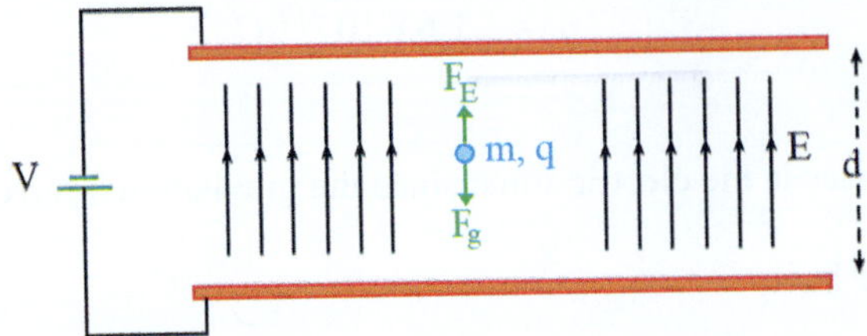

Fig. 2.11 A positively charged particle suspended in the air and in a uniform electric field

2.25. Based on the information given in the problem, we have:

$$\sigma = 10.6\ \mu C/m^2$$

$$\varepsilon_0 = 8.85 \times 10^{-12}\ \frac{C^2}{N.m^2}$$

The electric filed on the top and bottom of a very large surface with the surface charge density of σ can be calculated as follows (see Fig. 2.12).

$$E = \frac{\sigma}{2\varepsilon_0}$$

$$\Rightarrow E = \frac{10.6 \times 10^{-6}}{2 \times 8.85 \times 10^{-12}}$$

$$\Rightarrow E = 6 \times 10^5\ N/C$$

Choice (2) is the answer.

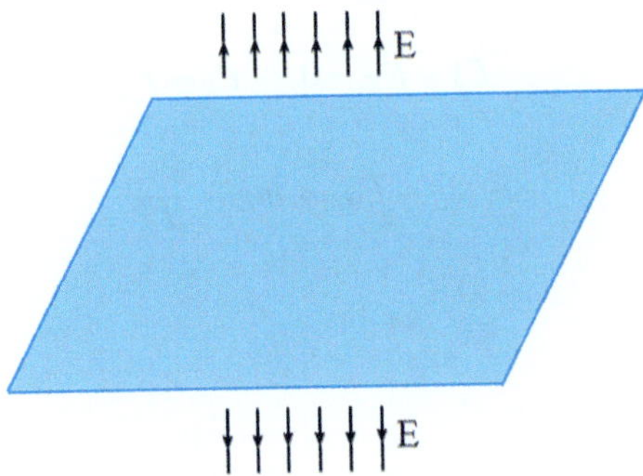

Fig. 2.12 A very large surface with the surface charge density

2.4 Electric Potential

2.26. Based on the information given in the problem, we have:

$$\Delta V = 3000\ V$$

$$E = 0.6\ mJ$$

The relation between the amount of energy or work and electric potential difference is as follows.

$$E = W = q\Delta V$$

Therefore:

$$q = \frac{W}{\Delta V} = \frac{0.6 \times 10^{-3}}{3000}$$

$$\Rightarrow q = 0.2\ \mu C$$

Choice (2) is the answer.

2.27. Based on the information given in the problem, we have:

$$E = 3000\ \frac{N}{C}$$

$$d = 0.02\ m$$

The relation between the electric potential difference of two points (V_{12}) and the distance between them in a uniform electric field is as follows.

$$E = \frac{\Delta V}{\Delta d} = \frac{V_1 - V_2}{\Delta d} = \frac{V_{12}}{\Delta d}$$

Therefore:

$$V_{12} = E \times \Delta d = 3000 \times 0.02$$

$$V_{12} = 60\ V$$

Choice (4) is the answer (Fig. 2.13).

Fig. 2.13 A uniform electric field

2.28. As we know, the electric potential around a point charge (q) can be calculated as follows.

$$V = k\frac{q}{r}$$

Herein, r is the radial distance from the point charge and $k = \frac{1}{4\pi\varepsilon_0} = 9 \times 10^9 \frac{N.m^2}{C^2}$ is the Coulomb constant. Moreover, ε_0 is the permittivity of free space which is equal to $8.85 \times 10^{-12} \frac{C^2}{N.m^2}$.

Hence:

$$V_A = k\frac{q}{1} = kq$$

$$V_B = k\frac{q}{2} = \frac{1}{2}kq$$

Therefore:

$$V_A = 2V_B$$

Choice (2) is the answer (Fig. 2.14).

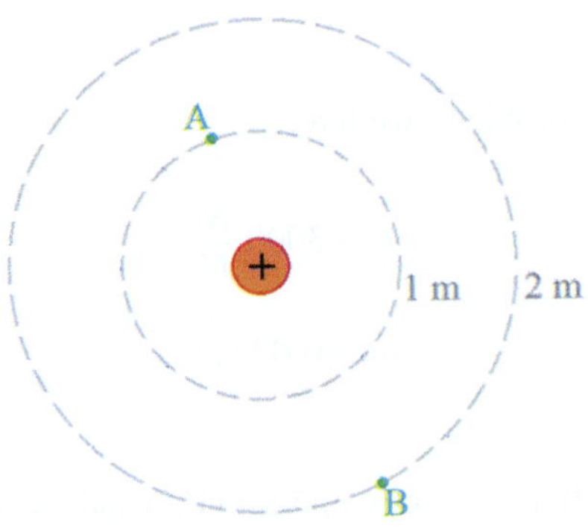

Fig. 2.14 The equipotential lines around a positive charge

2.30. Based on the information given in the problem, we have:

$$r' = 3r$$

$$q' = 3q$$

As we know, the electric potential around a point charge (q) at the radial distance r can be calculated as follows.

$$V = k\frac{q}{r}$$

where, $k = \frac{1}{4\pi\varepsilon_0} = 9 \times 10^9 \frac{N.m^2}{C^2}$ is the Coulomb constant. Herein, ε_0 is the permittivity of free space which is equal to $8.85 \times 10^{-12} \frac{C^2}{N.m^2}$.

The magnitude of electric filed in the second case can be calculated as follows.

$$V' = k\frac{q'}{r'}$$

$$\Rightarrow V' = k\frac{3q}{3r} = k\frac{q}{r}$$

$$\Rightarrow V' = V$$

$$\Rightarrow \frac{V'}{V} = 1$$

Choice (2) is the answer.

2.31. Based on the information given in the problem, we have:

$$q = 1\ \mu C$$

$$k = \frac{1}{4\pi\varepsilon_0} = 9 \times 10^9\ \frac{N.m^2}{C^2}$$

As we know, electric potential is a scalar quantity. Thus, the total electric potential at point p is the algebraic sum of the electric potentials.

The electric potential of each of the charges can be calculated as follows.

$$V_1 = V_2 = k\frac{q}{r}$$

$$\Rightarrow V_1 = V_2 = 9 \times 10^9 \times \frac{10^{-6}}{\sqrt{3^2 + 4^2}}$$

$$\Rightarrow V_1 = V_2 = 1800\ V$$

$$\Rightarrow V = V_1 + V_2 = 1800 + 1800$$

$$\Rightarrow V = 3600\ V$$

Choice (4) is the answer (Fig. 2.15).

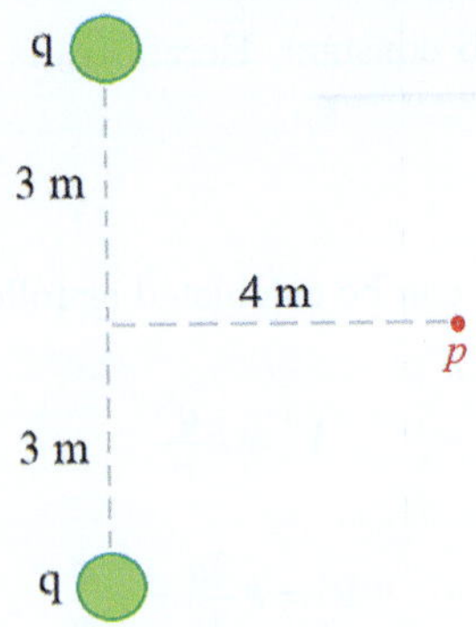

Fig. 2.15 Two charges and a point on a plane

2.33. Based on the information given in the problem, we have:

$$a = \sqrt{2}\, m$$

$$q_1 = 12\, nC$$

$$q_2 = -24\, nC$$

$$q_3 = 31\, nC$$

$$q_4 = 17\, nC$$

$$k = \frac{1}{4\pi\varepsilon_0} = 9 \times 10^9\, \frac{N.m^2}{C^2}$$

As we know, electric potential is a scalar quantity. Therefore, the total electric potential at a given point is the algebraic sum of the electric potentials. Hence, by using Fig. 2.16, we have:

$$V = k\frac{q_1}{r_1} + k\frac{q_2}{r_2} + k\frac{q_3}{r_3} + k\frac{q_4}{r_4}$$

The diagonal of the square can be calculated as follows.

$$d = a\sqrt{2} = \sqrt{2} \times \sqrt{2} = 2\, m$$

Therefore:

$$V = k\frac{q_1}{\frac{d}{2}} + k\frac{q_2}{\frac{d}{2}} + k\frac{q_3}{\frac{d}{2}} + k\frac{q_4}{\frac{d}{2}} = \frac{2k}{d}(q_1 + q_2 + q_3 + q_4)$$

$$\Rightarrow V = \frac{2 \times 9 \times 10^9}{2}(-9 + 2 + 3 + 4) = 9 \times 10^9 \times 0$$

$$\Rightarrow V = 0\, V$$

Choice (4) is the answer.

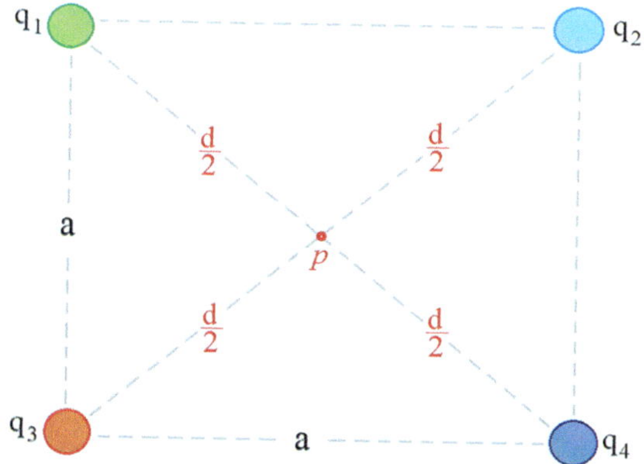

Fig. 2.16 Four charges on the corners (vertices) of a square

2.36. As is shown in Fig. 2.17, the equipotential surfaces around a point charge are the spherical surfaces in 3D and circular lines in 2D. As can be noticed, the electric potential of equipotential surfaces around a positive point charge is positive. Moreover, the electric potential of equipotential surfaces around a negative point charge is negative. Choice (4) is the answer.

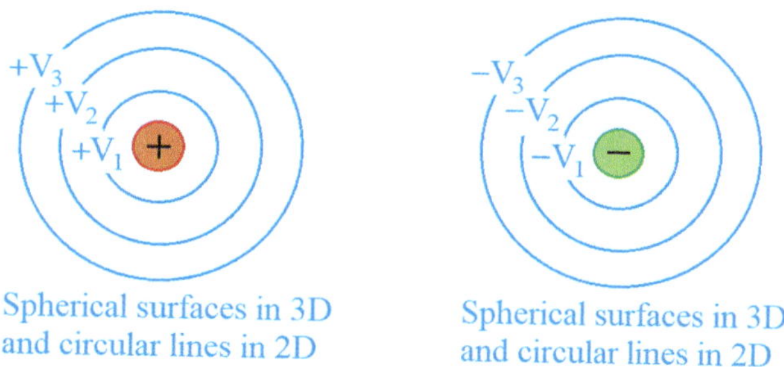

Fig. 2.17 The equipotential surfaces around a positive and negative point charges in 3D and 2D

2.37. Based on the information given in the problem, we have:

$$k = \frac{1}{4\pi\varepsilon_0} = 9 \times 10^9 \ \frac{N.m^2}{C^2}$$

The electric potential at points A and B can be calculated as follows.

$$V_A = k\frac{q}{d} + k\frac{2q}{4d} = \frac{3}{2}\frac{kq}{d}$$

$$V_B = k\frac{q}{4d} + k\frac{2q}{d} = \frac{9}{4}\frac{kq}{d}$$

$$\Rightarrow \Delta V = \frac{9}{4}\frac{kq}{d} - \frac{3}{2}\frac{kq}{d} = \frac{3}{4}\frac{kq}{d}$$

$$\Rightarrow \Delta V = \frac{3q}{16\pi\varepsilon_0 d}$$

Choice (2) is the answer (Fig. 2.18).

Fig. 2.18 Two point charges and two points on the corners (vertices) of a rectangle

2.5 Electric Potential Energy

2.38. Based on the information given in the problem, we have:

$$q_1 = 1\ \mu C$$

$$q_2 = 1\ \mu C$$

$$r = 2\ m$$

$$k = \frac{1}{4\pi\varepsilon_0} = 9 \times 10^9\ \frac{N.m^2}{C^2}$$

The electric potential energy of a system of two charges can be calculated as follows.

$$U = k\frac{q_1 q_2}{r}$$

Therefore:

$$U = \frac{9 \times 10^9 \times \left(1 \times 10^{-6}\right)^2}{2}$$

$$\Rightarrow U = 4.5\ mJ$$

Choice (4) is the answer.

2.39. Based on the information given in the problem, we have:

$$d = 12\ cm$$

$$q_1 = 15\ \mu C$$

$$q_2 = -60\ \mu C$$

$$q_3 = 30\ \mu C$$

$$k = \frac{1}{4\pi\varepsilon_0} = 9 \times 10^9\ \frac{N.m^2}{C^2}$$

The electric potential energy of a system of three charges can be calculated as follows.

$$U = U_{12} + U_{13} + U_{23}$$

$$\Rightarrow U = k\frac{q_1 q_2}{r_{12}} + k\frac{q_1 q_3}{r_{13}} + k\frac{q_2 q_3}{r_{23}}$$

$$\Rightarrow U = 9 \times 10^9 \left[\frac{(15 \times 10^{-6})(-60 \times 10^{-6})}{0.12} + \frac{(15 \times 10^{-6})(30 \times 10^{-6})}{0.12} + \frac{(-60 \times 10^{-6})(30 \times 10^{-6})}{0.12}\right]$$

$$\Rightarrow U = \frac{9 \times 10^9}{0.12}[-900 + 450 - 1800] \times 10^{-12}$$

$$\Rightarrow U = -168.75\ J$$

Choice (1) is the answer (Fig. 2.19).

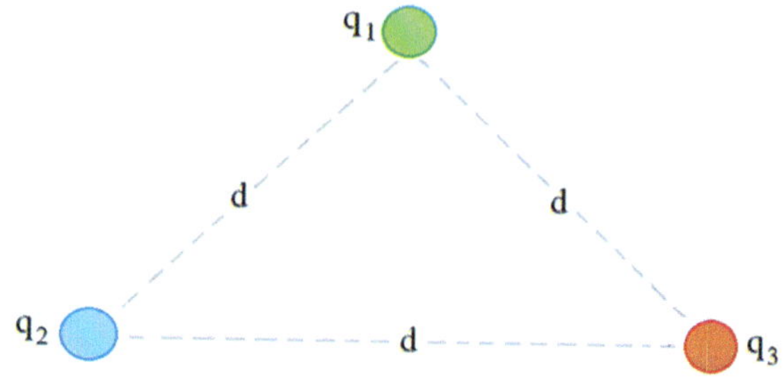

Fig. 2.19 Three charges on the corners (vertices) of an equilateral triangle

2.41. The amount of work is irrespective of the path. Hence, the amount of work needed to move charge q' from point A to point B can be calculated as follows.

$$W = \Delta U$$

$$\Rightarrow W = q'\Delta V = q'(V_B - V_A)$$

The electric potential at points A and B can be calculated as follows.

$$V_A = k\frac{q'q}{a} + k\frac{q'q}{a\sqrt{2}}$$

$$V_B = k\frac{q'q}{a\sqrt{2}} + k\frac{q'q}{a}$$

As can be seen, $V_A = V_B$. Therefore:

$$W = 0$$

Choice (1) is the answer (Fig. 2.20).

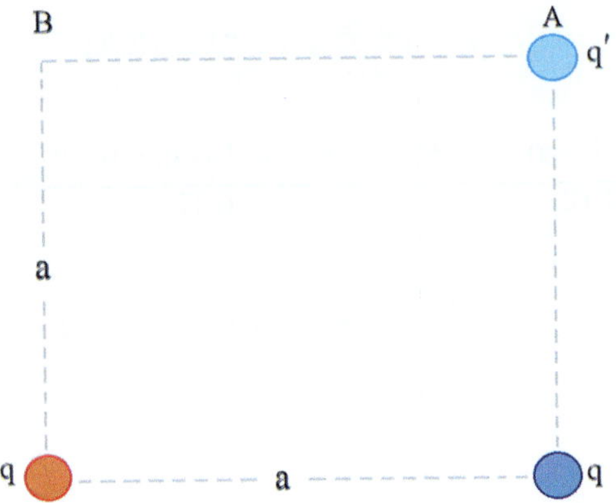

Fig. 2.20 Three point charges and one point on the corners (vertices) of a square

2.6 Electric Energy and Electric Energy Density

2.42. Based on the information given in the problem, we have:

$$E = 10^3\ N/C$$

$$\varepsilon = \varepsilon_0 = 8.85 \times 10^{-12} \frac{C^2}{N.m^2}$$

As we know, the electric energy density (J/m^3) in a space can be calculated as follows.

$$\omega = \frac{1}{2}\varepsilon E^2$$

Therefore:

$$\Rightarrow \omega = \frac{1}{2} \times 8.85 \times 10^{-12} \times \left(10^3\right)^2$$

$$\Rightarrow \omega = 4.42 \times 10^{-6}\ J/m^3 = 4.42\ \mu J/m^3$$

Choice (4) is the answer.

2.43. Based on the information given in the problem, we have:

$$a \times b \times c = 0.02 \times 0.01 \times 0.002\ m$$

$$\omega = 4.42\ \mu J/m^3$$

The total electric energy in a space can be calculated as follows.

$$W = \omega V$$

Hence:

$$W = \omega \times a \times b \times c$$

$$\Rightarrow W = 4.42 \times 10^{-6} \times 0.02 \times 0.01 \times 0.002$$

$$\Rightarrow W = 17.68 \times 10^{-13}\ J = 1.768\ pJ$$

Choice (4) is the answer.

References

1. Rahmani-Andebili, M., General Physics I - Practice Problems, Methods, and Solutions, Springer Nature, 2025.
2. Rahmani-Andebili, M., Calculus III – Practice Problems, Methods, and Solutions, Springer Nature, 2023.
3. Rahmani-Andebili, M., Calculus II – Practice Problems, Methods, and Solutions, Springer Nature, 2023.
4. Rahmani-Andebili, M., Calculus I (2nd Ed.) – Practice Problems, Methods, and Solutions, Springer Nature, 2023.
5. Rahmani-Andebili, M., Precalculus (2nd Ed.) – Practice Problems, Methods, and Solutions, Springer Nature, 2024.

3 Electrical Charge, Capacitance, Capacitor, Current, Resistance, and Resistor: Part A

Abstract

In this chapter, the basic and advanced problems of Electrical Charge, Capacitance, Capacitor, Current, Resistance, and Resistor are studied. The subjects include Capacitance and Equivalent Capacitance of a Circuit, Electrical Charge and Voltage of Capacitors, Electrical Energy and Energy Density Stored in Capacitors, Resistance and Equivalent Resistance of a Circuit, Ohm's Law, Electrical Current and Voltage of Resistors, and Electrical Power and Energy Wasted in Resistors. Herein, different types of problems and exercises are presented that are categorized as follows.

- ***Problems with detailed solution***: They have been designed to teach students the subjects in detail. Moreover, they have been categorized in different levels based on their difficulty levels (easy, normal, and hard) and calculation amounts (small, normal, and large).
- ***Partially solved exercises***: They have been designed to encourage students to practice problems while guiding them through the problem-solving procedure and hinting the required formulas.
- ***Exercises with final answer***: They have been designed to encourage students to practice more by themselves while hinting them by the final answer as well as to help instructors to give tests or quizzes.

3.1 Capacitance and Equivalent Capacitance of a Circuit

Problem

3.1. The surface area of each plate of a flat capacitor is A. Moreover, the distance between its plates is d where air fills the gap. If a conductor fills 50% of the gap, calculate the capacitance of the capacitor [1–5] (Fig. 3.1).

Difficulty level ○ Easy ● Normal ○ Hard 

Calculation amount ● Small ○ Normal ○ Large

1) $\frac{\varepsilon_0 A}{d}$
2) $\frac{2\varepsilon_0 A}{d}$
3) $\frac{\varepsilon_0 A}{2d}$
4) $\frac{4\varepsilon_0 A}{d}$

M. Rahmani-Andebili, *General Physics II*, https://doi.org/10.1007/978-3-031-92866-6_3

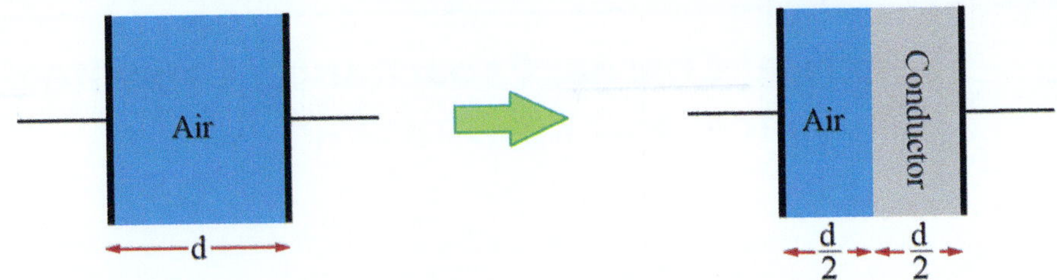

Fig. 3.1 A flat capacitor

Problem

3.2. In Fig. 3.2, calculate the capacitance of the capacitor. Herein, the surface area of each plate of the flat capacitor is A.

Difficulty level ○ Easy ● Normal ○ Hard

Calculation amount ● Small ○ Normal ○ Large

1) $\frac{(\varepsilon_1 + \varepsilon_2)A}{2d}$
2) $\frac{(\varepsilon_1 + \varepsilon_2)A}{d}$
3) $\frac{(\varepsilon_1 + \varepsilon_2)A}{\varepsilon_1 \varepsilon_2 d}$
4) $\frac{\varepsilon_1 \varepsilon_2 A}{2(\varepsilon_1 + \varepsilon_2)d}$

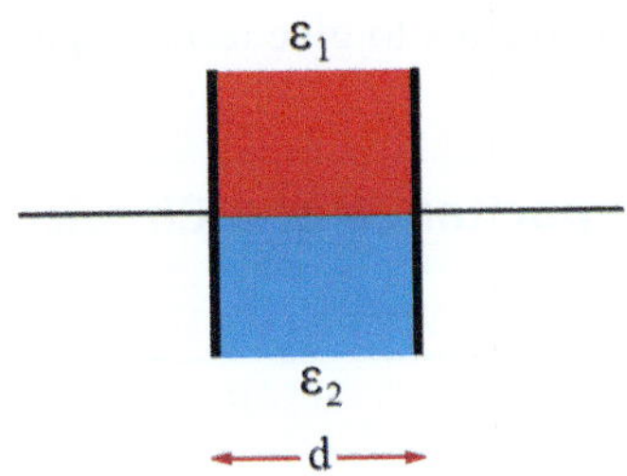

Fig. 3.2 A flat capacitor including two different electrical permittivity

Problem

3.3. In Fig. 3.3, calculate the capacitance of the capacitor. Herein, the surface area of each plate of the flat capacitor is A.

Difficulty level ○ Easy ● Normal ○ Hard

Calculation amount ○ Small ● Normal ○ Large

1) $\frac{2(\varepsilon_1 + \varepsilon_2)A}{d}$
2) $\frac{(\varepsilon_1 + \varepsilon_2)A}{d}$

3) $\dfrac{\varepsilon_1\varepsilon_2 A}{(\varepsilon_1+\varepsilon_2)d}$

4) $\dfrac{2\varepsilon_1\varepsilon_2 A}{(\varepsilon_1+\varepsilon_2)d}$

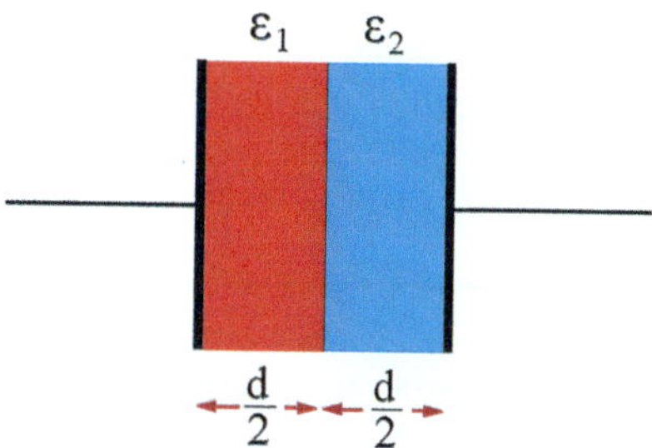

Fig. 3.3 A flat capacitor including two different electrical permittivity

Problem

3.4. The radius of circular plates of a flat capacitor is 1 *cm* and the distance between its plates is 1 *mm* filled by air. Calculate the capacitance of the capacitor. The permittivity of free space is $\varepsilon_0 = 8.85 \times 10^{-12} \frac{C^2}{N.m^2}$.

Difficulty level ● Easy ○ Normal ○ Hard

Calculation amount ● Small ○ Normal ○ Large

1) 2.78 *mF*
2) 2.78 *μF*
3) 2.78 *nF*
4) 2.78 *pF*

Problem

3.5. In Problem 3.4, if the capacitor is connected to a 120 *V* voltage source, calculate the charge stored in it.

Difficulty level ● Easy ○ Normal ○ Hard

Calculation amount ● Small ○ Normal ○ Large

1) 43.16×10^{-12} C
2) 333.6×10^{-12} C
3) 43.1×10^{-9} C
4) 0.023×10^{-9} C

Problem

3.6. In the circuit shown in Fig. 3.4, the capacitors are connected in series. Calculate the equivalent capacitance of the circuit.

Difficulty level ● Easy ○ Normal ○ Hard

Calculation amount ● Small ○ Normal ○ Large

1) 11 F
2) 1 F
3) 4 F
4) 2 F

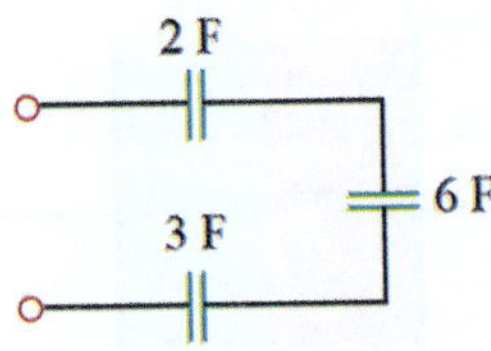

Fig. 3.4 The series connection of capacitors

Problem

3.7. In the circuit shown in Fig. 3.5, the capacitors are connected in parallel. Calculate the equivalent capacitance of the circuit.

Difficulty level ● Easy ○ Normal ○ Hard
Calculation amount ● Small ○ Normal ○ Large

1) 11 F
2) 1 F
3) 4 F
4) 2 F

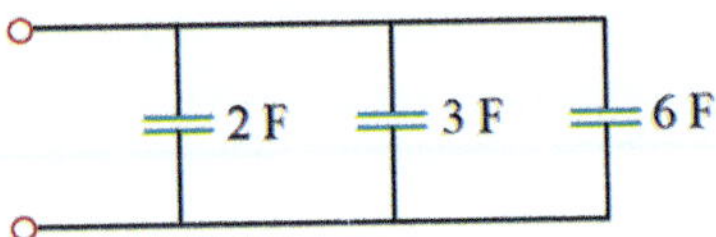

Fig. 3.5 The parallel connection of capacitors

Problem

3.8. In the circuit shown in Fig. 3.6, the capacitors are connected in series-parallel from. Calculate the equivalent capacitance of the circuit.

Difficulty level ● Easy ○ Normal ○ Hard
Calculation amount ● Small ○ Normal ○ Large

1) 4 F
2) 2.5 F
3) 2 F
4) 1 F

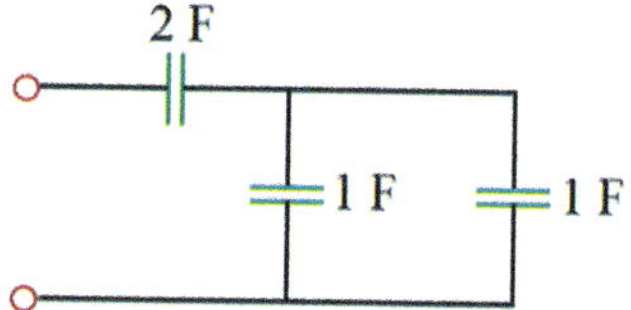

Fig. 3.6 The series-parallel connection of capacitors

Partially Solved Exercise

3.9. Calculate the equivalent capacitance of the circuit shown in Fig. 3.7.

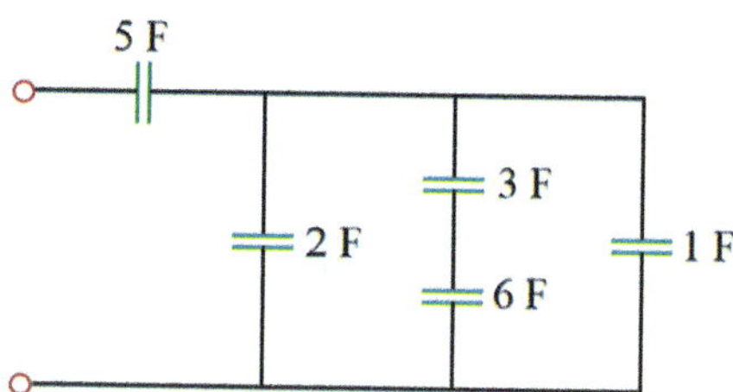

Fig. 3.7 The series-parallel connection of capacitors

Solution

First, the series connection of the middle branch should be simplified as follows.

$$\frac{1}{C_{3,6}} = \frac{1}{C_3} + \frac{1}{C_6}$$

Or:

$$C_{3,6} = \frac{C_3 C_6}{C_3 + C_6}$$

$$\Rightarrow C_{3,6} = \frac{(\quad)\times(\quad)}{(\quad)+(\quad)} = 2\,F$$

Then, the three parallel branches can be combined as follows.

$$C_{2,2,1} = C_2 + C_{3,6} + C_1 = (\quad)+(\quad)+(\quad) = 5\,F$$

Finally, two 5 F capacitors are left connected in series. Thus:

$$C_{eq} = \frac{C_5}{2} = \frac{(\quad)}{2}$$

$$\Rightarrow C_{eq} = 2.5\,F$$

Notes

In this problem, the relations below have been used.

The equivalent capacitance of n parallel capacitors can be calculated as follows.

$$C_{eq} = C_1 + \ldots + C_n$$

If $C_1 = \ldots = C_n = C$, we have:

$$C_{eq} = nC$$

The equivalent capacitance of n series capacitors can be calculated as follows.

$$\frac{1}{C_{eq}} = \frac{1}{C_1} + \ldots + \frac{1}{C_n}$$

If $C_1 = \ldots = C_n = C$, we have:

$$C_{eq} = \frac{C}{n}$$

If $n = 2$, we have:

$$\frac{1}{C_{eq}} = \frac{1}{C_1} + \frac{1}{C_2} \Rightarrow C_{eq} = \frac{C_1 C_2}{C_1 + C_2}$$

3.2 Electrical Charge and Voltage of Capacitors

Problem

3.10. Three similar capacitors are connected to a voltage source in two different tests. In the first one, they are connected in series, and in the second one in parallel. Calculate the ratio of total charge of the capacitors in the first test to the second one.

Difficulty level ○ Easy ● Normal ○ Hard 

Calculation amount ● Small ○ Normal ○ Large

1) $\frac{1}{9}$
2) $\frac{1}{3}$
3) 3
4) 9

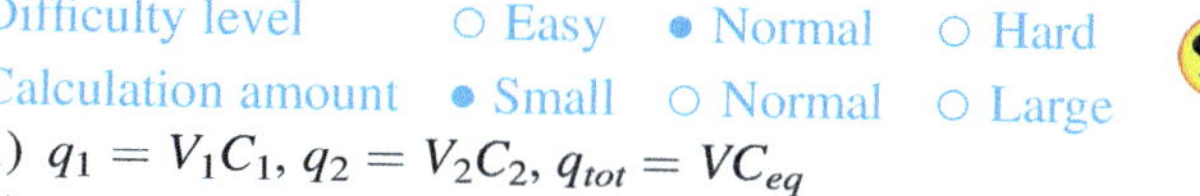

3.11. Which one of the following choices is wrong about the series circuit shown in Fig. 3.8.

Difficulty level ○ Easy ● Normal ○ Hard

Calculation amount ● Small ○ Normal ○ Large

1) $q_1 = V_1C_1,\ q_2 = V_2C_2,\ q_{tot} = VC_{eq}$
2) $q_1 = q_2 = q_{tot}$
3) $V_1 = \frac{C_2}{C_1 + C_2}V,\ V_2 = \frac{C_1}{C_1 + C_2}V$
4) $V_1 = V_2 = V$

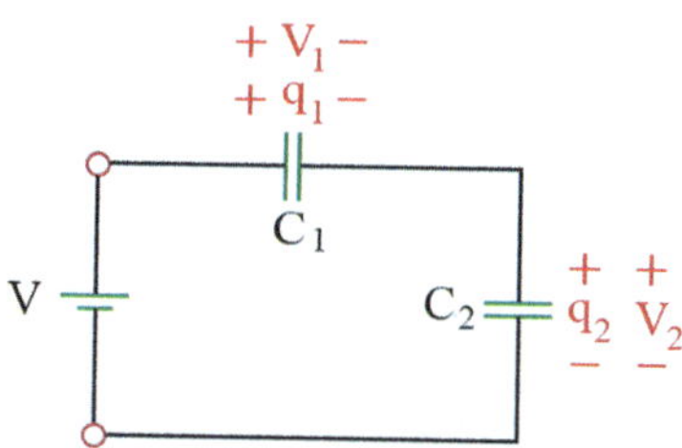

Fig. 3.8 The series connection of two capacitors

Problem

3.12. Which one of the following choices is wrong about a parallel circuit shown in Fig. 3.9.

Difficulty level ○ Easy ● Normal ○ Hard

Calculation amount ● Small ○ Normal

1) $q_1 = V_1C_1,\ q_2 = V_2C_2,\ q_{tot} = VC_{eq}$
2) $V_1 = V_2 = V$
3) $q_1 = \frac{C_1}{C_1 + C_2}q_{tot},\ q_2 = \frac{C_2}{C_1 + C_2}q_{tot}$
4) $q_1 = q_2 = q_{tot}$

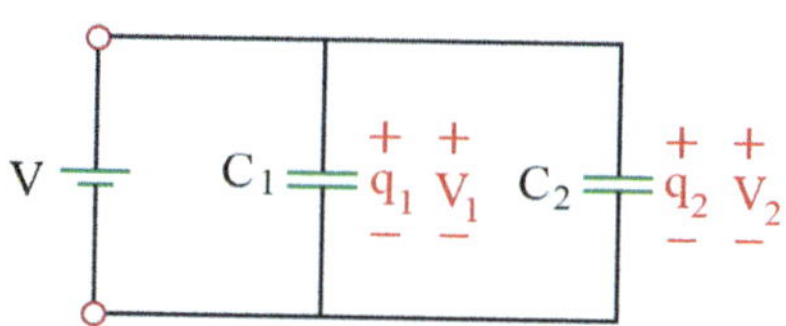

Fig. 3.9 The parallel connection of two capacitors

Problem

3.13. In the circuit below, $C_1 = 2\ \mu F$, $C_2 = 3\ \mu F$, $V = 150\ V$. Calculate the voltage across C_1 (Fig. 3.10).

Difficulty level ○ Easy ● Normal ○ Hard

Calculation amount ● Small ○ Normal ○ Large

1) 120 V
2) 60 V
3) 30 V
4) 240 V

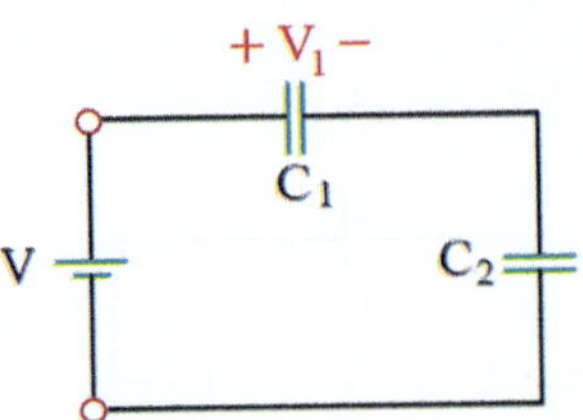

Fig. 3.10 The series connection of two capacitors

Problem

3.14. In the circuit below, $C_1 = 4\ \mu F$, $C_2 = 6\ \mu F$, $V = 200\ V$. Calculate the charge stored in C_1 (Fig. 3.11).

Difficulty level ○ Easy ● Normal ○ Hard

Calculation amount ○ Small ● Normal ○ Large

1) $4.8\times 10^{-4}\ C$
2) $48 \times 10^{-4}\ C$
3) $1.2 \times 10^{-4}\ C$
4) $12 \times 10^{-4}\ C$

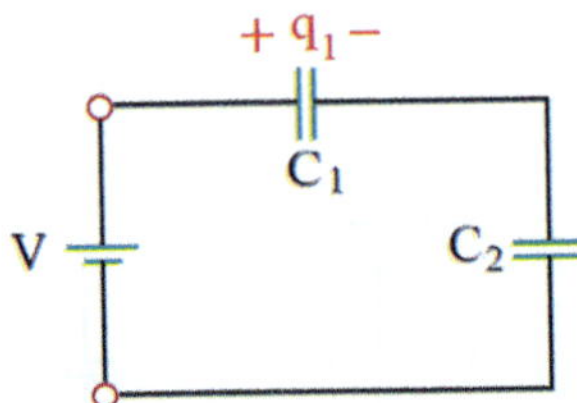

Fig. 3.11 The series connection of two capacitors

Problem

3.15. In the circuit below, $C_1 = 2\ \mu F$, $C_2 = 3\ \mu F$, $V = 150\ V$. Calculate the voltage across C_1 (Fig. 3.12).

Difficulty level ○ Easy ● Normal ○ Hard
Calculation amount ● Small ○ Normal ○ Large

1) 90 V
2) 60 V
3) 30 V
4) 150 V

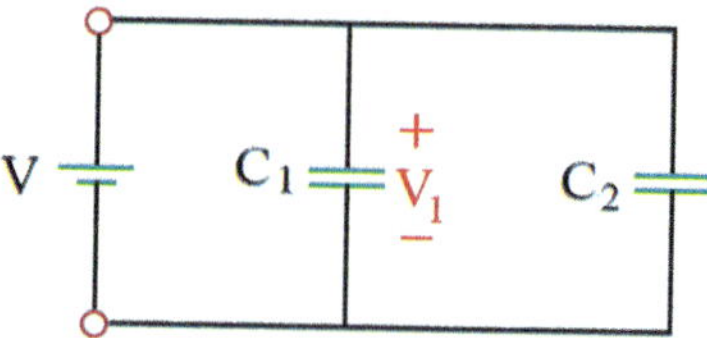

Fig. 3.12 The parallel connection of two capacitors

Problem

3.16. In the circuit below, $C_1 = 4\ \mu F$, $C_2 = 6\ \mu F$, $V = 200\ V$. Calculate the charge stored in C_1 (Fig. 3.13).

Difficulty level ○ Easy ● Normal ○ Hard
Calculation amount ○ Small ● Normal ○ Large

1) $4 \times 10^{-4}\ C$
2) $6 \times 10^{-4}\ C$
3) $8 \times 10^{-4}\ C$
4) $12 \times 10^{-4}\ C$

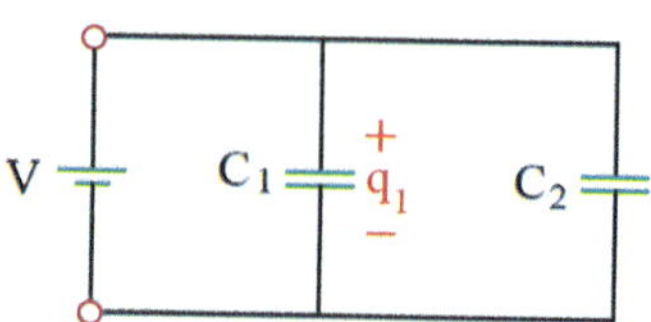

Fig. 3.13 The parallel connection of two capacitors

Problem

3.17. A capacitor with the capacity of 2 μF is charged by a battery with the voltage of 100 V and then is separated from it. This capacitor is connected to a charge-free 3 μF capacitor. In this condition, calculate the voltage of each capacitor.

Difficulty level ○ Easy ○ Normal ● Hard

Calculation amount ○ Small ● Normal ○ Large

1) 20 V
2) 0 V
3) 40 V
4) 100 V

Problem

3.18. While the capacitor is connected to a battery, the distance between the plates of the capacitor is increased. Which one of the following choices is correct.

Difficulty level ○ Easy ○ Normal ● Hard

Calculation amount ● Small ○ Normal ○ Large

1) V remains constant, C decreases, q increases
2) V increases, C increases, q increases
3) V increases, C decreases, q decreases
4) V remains constant, C decreases, q decreases

Problem

3.19. A capacitor is fully charged and then disconnected from its battery. After that, the distance between the plates of the capacitor is increased. Which one of the following choices is correct.

Difficulty level ○ Easy ○ Normal ● Hard

Calculation amount ● Small ○ Normal ○ Large

1) V remains constant, C decreases, q increases
2) V increases, C increases, q remains constant
3) V increases, C decreases, q remains constant
4) V remains constant, C decreases, q decreases

3.3 Electrical Energy Stored in Capacitors

Problem

3.20. Three capacitors, that is, $C_1 = 1.6\ \mu F$, $C_2 = 2\ \mu F$, and $C_3 = 4\ \mu F$ are connected in parallel to a voltage source. Calculate the electrical energy stored in C_3 if $q_1 = 400\ \mu C$.

Difficulty level ○ Easy ● Normal ○ Hard

Calculation amount ○ Small ● Normal ○ Large

1) 40 mJ
2) 0.4 mJ
3) 40 J
4) 0.4 J

Partially Solved Exercise

3.21. Calculate the total electrical energy stored in three capacitors, that is, $C_1 = 1\ \mu F$, $C_2 = 2\ \mu F$, and $C_3 = 2\ \mu F$ that are connected in parallel to a 1000 V voltage source.

Solution

As we know, the electrical energy stored in a capacitor can be calculated by using one of the following relations.

$$W = \frac{1}{2}CV^2 = \frac{1}{2}qV = \frac{1}{2}\frac{q^2}{C}$$

Since the capacitors have been connected in parallel, their equivalent capacitance can be calculated as follows.

$$C_{eq} = (\quad) + (\quad) + (\quad) = 5\ \mu F = 5 \times 10^{-6}\ F$$

Hence, the total electrical energy stored in capacitors is as follows.

$$W_{tot} = \frac{1}{2}C_{eq}V^2 = \frac{1}{2} \times (\qquad\qquad) \times (\qquad)^2$$

$$\Rightarrow W_{tot} = 2.5\ J$$

Partially Solved Exercise

3.22. Calculate the total electrical energy stored in three capacitors, that is, $C_1 = 1\ \mu F$, $C_2 = 2\ \mu F$, and $C_3 = 2\ \mu F$ that are connected in series to a 1000 V voltage source.

Solution

As we know, the electrical energy stored in a capacitor can be calculated by using one of the following relations.

$$W = \frac{1}{2}CV^2 = \frac{1}{2}qV = \frac{1}{2}\frac{q^2}{C}$$

Since the capacitors have been connected in parallel, their equivalent capacitance can be calculated as follows.

$$\frac{1}{C_{eq}} = \frac{1}{(\quad)} + \frac{1}{(\quad)} + \frac{1}{(\quad)} = \frac{(\quad) + (\quad) + (\quad)}{(\quad)} = 2$$

$$C_{eq} = 0.5\ \mu F = 0.5 \times 10^{-6}\ F$$

Hence, the total electrical energy stored in capacitors is as follows.

$$W_{tot} = \frac{1}{2}C_{eq}V^2 = \frac{1}{2} \times (\qquad\qquad) \times (\qquad)^2$$

$$\Rightarrow W_{tot} = 0.25\ J$$

3.4 Resistance and Equivalent Resistance of a Circuit

Problem

3.23. The diameter of the circular cross-sectional area, length, and resistance of a wire are 1 mm, 2 m, and 0.1 Ω, respectively. Calculate the special resistance (resistivity) of the wire.

Difficulty level ○ Easy ● Normal ○ Hard

Calculation amount ○ Small ● Normal ○ Large

1) $3.92 \times 10^{-8}\ \Omega m$
2) $3.92 \times 10^{-6}\ \Omega m$
3) $15.68 \times 10^{-8}\ \Omega m$
4) $15.68 \times 10^{-6}\ \Omega m$

Problem

3.24. Consider two wires. The length of the first one is twice the length of the second wire. Moreover, the diameter of circular cross-sectional area of the first wire is half of the one of the second wire. In addition, the special resistance (resistivity) of the first wire is one-third of the one of the second wire. Calculate the ratio of resistance of the first wire to the second wire.

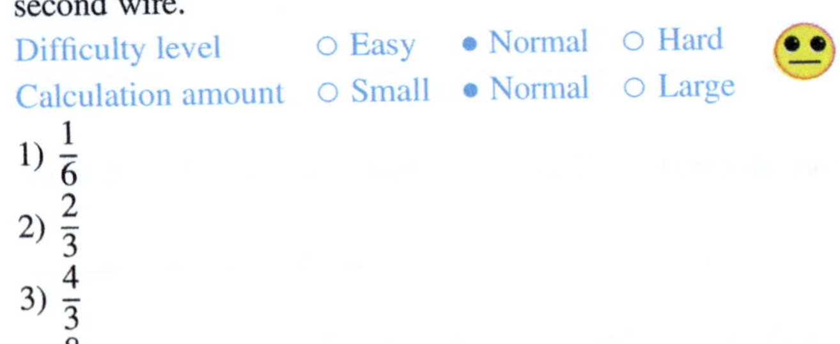

Difficulty level ○ Easy ● Normal ○ Hard

Calculation amount ○ Small ● Normal ○ Large

1) $\frac{1}{6}$
2) $\frac{2}{3}$
3) $\frac{4}{3}$
4) $\frac{8}{3}$

Problem

3.25. In the circuit shown in Fig. 3.14, the resistors are connected in series. Calculate the equivalent resistance of the circuit.

Difficulty level ● Easy ○ Normal ○ Hard

Calculation amount ● Small ○ Normal ○ Large

1) 1 Ω
2) 11 Ω
3) 10 Ω
4) 4 Ω

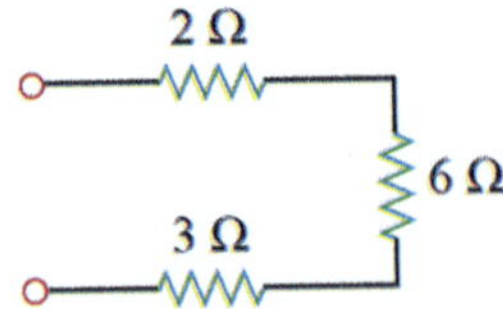

Fig. 3.14 The series connection of resistors

Problem

3.26. In the circuit shown in Fig. 3.15, the resistors are connected in parallel. Calculate the equivalent resistance of the circuit.

Difficulty level ● Easy ○ Normal ○ Hard
Calculation amount ● Small ○ Normal ○ Large

1) 1 Ω
2) 11 Ω
3) 10 Ω
4) 4 Ω

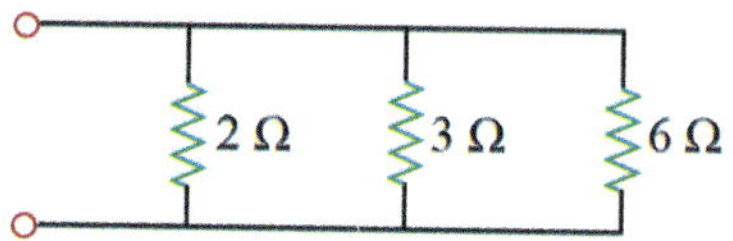

Fig. 3.15 The parallel connection of resistors

Problem

3.27. In the circuit shown in Fig. 3.16, the resistors are connected in series-parallel from. Calculate the equivalent resistance of the circuit.

Difficulty level ● Easy ○ Normal ○ Hard
Calculation amount ● Small ○ Normal ○ Large

1) 4 Ω
2) 1 Ω
3) 0.4 Ω
4) 2.5 Ω

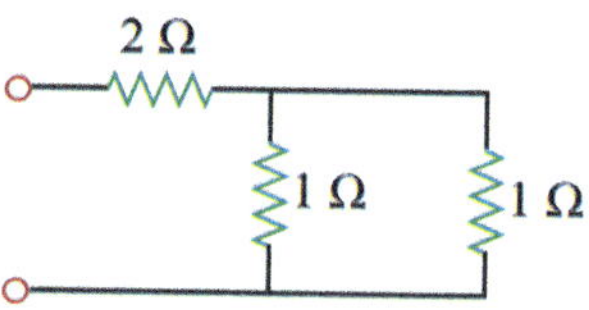

Fig. 3.16 The series-parallel connection of resistors

Partially Solved Exercise

3.28. Calculate the equivalent resistance of the circuit shown in Fig. 3.17.

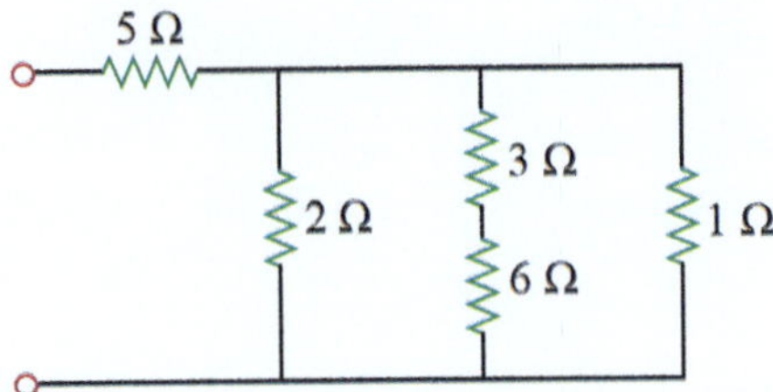

Fig. 3.17 The series-parallel connection of resistors

Solution

The equivalent resistance of the circuit can be calculated as follows. First, the series connection of the middle branch should be simplified as follows.

$$R_{3,6} = R_3 + R_6 = (\quad) + (\quad) = 9\ \Omega$$

Then, the three parallel branches can be combined as follows.

$$\frac{1}{R_{2,9,1}} = \frac{1}{R_2} + \frac{1}{R_9} + \frac{1}{R_1}$$

$$\Rightarrow \frac{1}{R_{2,9,1}} = \frac{1}{(\quad)} + \frac{1}{(\quad)} + \frac{1}{(\quad)} = \frac{(\quad) + (\quad) + (\quad)}{18} = \frac{29}{18}$$

$$\Rightarrow R_{2,9,1} = \frac{18}{29}\ \Omega$$

Finally, there are only two resistors connected in series. Thus:

$$R_{eq} = (\quad) + (\quad) = \frac{145 + 18}{29}$$

$$\Rightarrow R_{eq} = \frac{163}{29}\ \Omega$$

Notes

In this problem, the relations below have been used.

The equivalent resistance of n series resistors can be calculated as follows.

$$R_{eq} = R_1 + \ldots + R_n$$

If $R_1 = \ldots = R_n = R$, we have:

$$R_{eq} = nR$$

The equivalent resistance of n parallel resistors can be calculated as follows.

$$\frac{1}{R_{eq}} = \frac{1}{R_1} + \ldots + \frac{1}{R_n}$$

If $R_1 = \ldots = R_n = R$, we have:

$$R_{eq} = \frac{R}{n}$$

If $n = 2$, we have:

$$\frac{1}{R_{eq}} = \frac{1}{R_1} + \frac{1}{R_2} \Rightarrow R_{eq} = \frac{R_1 R_2}{R_1 + R_2}$$

3.5 Ohm's Law and Electrical Current and Voltage of Resistors

Problem

3.29. Which one of the following choices is wrong about a series circuit shown in Fig. 3.18.

Difficulty level ○ Easy ● Normal ○ Hard 

Calculation amount ● Small ○ Normal ○ Large

1) $I_1 = \frac{V_1}{R_1}, I_2 = \frac{V_2}{R_2}, I = \frac{V}{R_{eq}}$

2) $I_1 = I_2 = I$

3) $V_1 = \frac{R_1}{R_1 + R_2} V, V_2 = \frac{R_2}{R_1 + R_2} V$

4) $V_1 = V_2 = V$

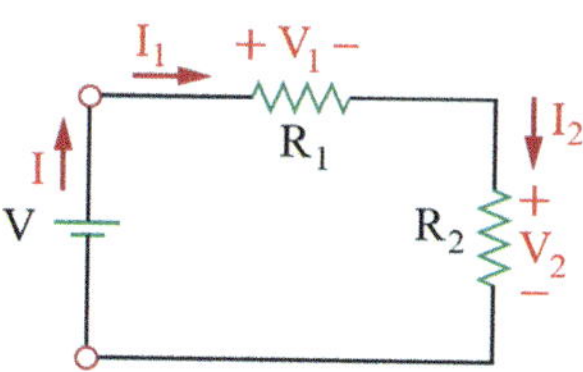

Fig. 3.18 The series connection of two resistors

Problem

3.30. Which one of the following choices is wrong about a parallel circuit shown in Fig. 3.19.

Difficulty level ○ Easy ● Normal ○ Hard

Calculation amount ● Small ○ Normal ○ Large

1) $V_1 = I_1R_1$, $V_2 = I_2R_2$, $V = IR_{eq}$
2) $V_1 = V_2 = V$
3) $I_1 = \frac{R_2}{R_1 + R_2}I, I_2 = \frac{R_1}{R_1 + R_2}I$
4) $I_1 = I_2 = I$

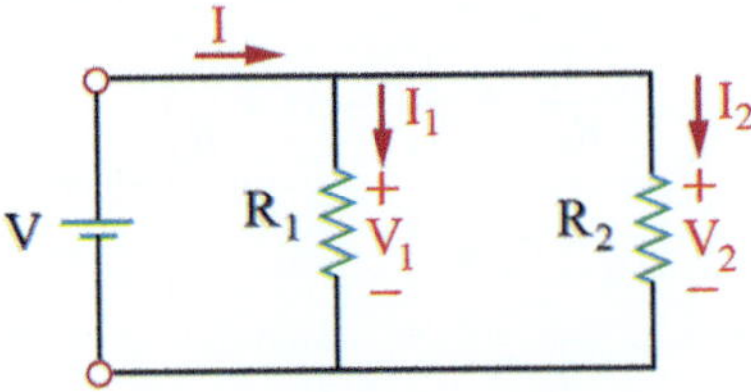

Fig. 3.19 The parallel connection of two resistors

Problem

3.31. In the circuit shown in Fig. 3.20, calculate the current of each resistor.

Difficulty level ● Easy ○ Normal ○ Hard

Calculation amount ● Small ○ Normal ○ Large

1) $I_1 = 10\ A$, $I_2 = 20\ A$
2) $I_1 = 20\ A$, $I_2 = 10\ A$
3) $I_1 = 20\ A$, $I_2 = 20\ A$
4) $I_1 = 10\ A$, $I_2 = 10\ A$

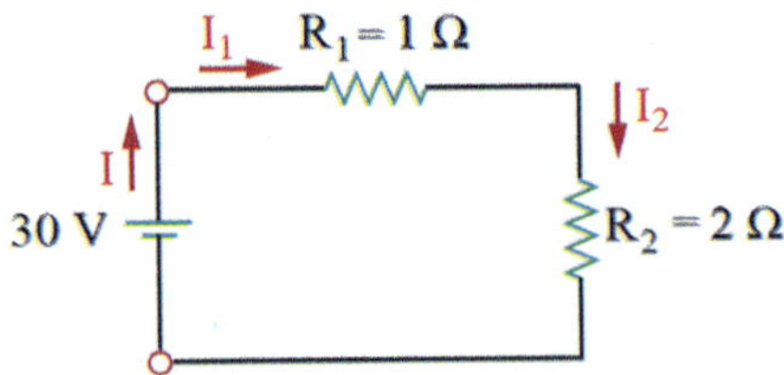

Fig. 3.20 The series connection of two resistors

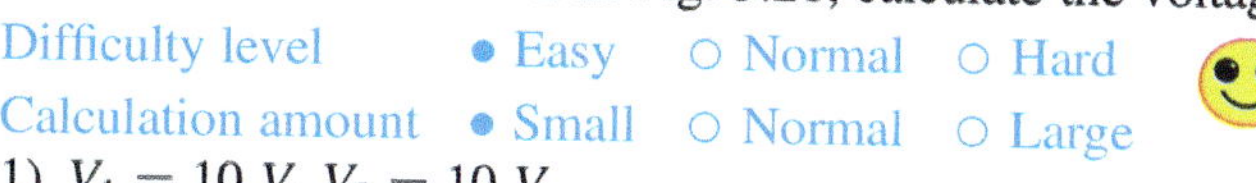

3.32. In the circuit shown in Fig. 3.21, calculate the voltage of each resistor.

Difficulty level ● Easy ○ Normal ○ Hard

Calculation amount ● Small ○ Normal ○ Large

1) $V_1 = 10\ V, V_2 = 10\ V$
2) $V_1 = 20\ V, V_2 = 20\ V$
3) $V_1 = 10\ V, V_2 = 20\ V$
4) $V_1 = 20\ V, V_2 = 20\ V$

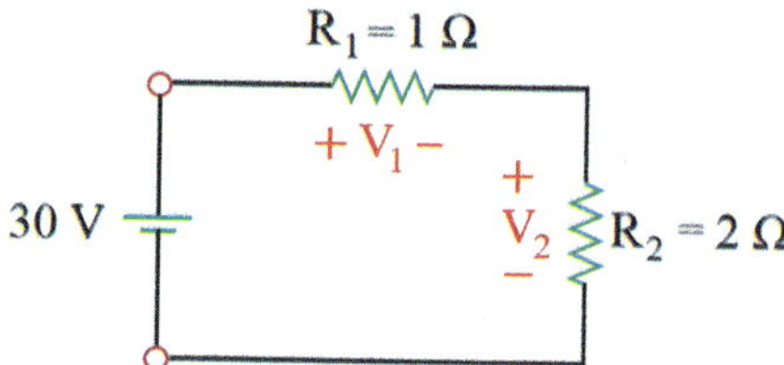

Fig. 3.21 The series connection of two resistors

Exercise

3.33. Calculate the voltage of the third resistor in the circuit shown in Fig. 3.22.

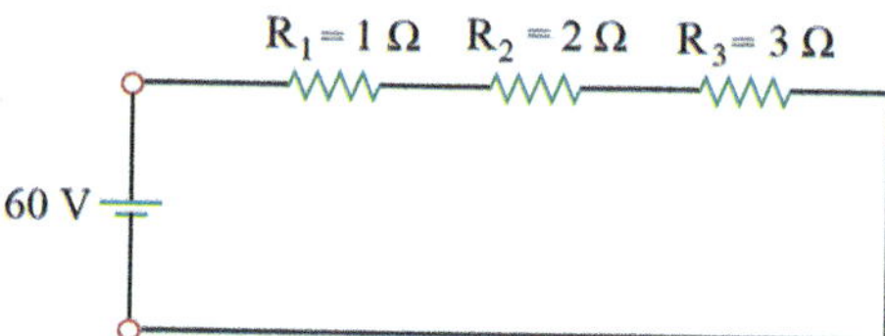

Fig. 3.22 The series connection of three resistors

Final Answer

$V_3 = 30\ V$

Problem

3.34. In the circuit shown in Fig. 3.23, calculate the voltage of each resistor.

Difficulty level ● Easy ○ Normal ○ Hard

Calculation amount ● Small ○ Normal ○ Large

1) $V_1 = 30\ V, V_2 = 30\ V$
2) $V_1 = 10\ V, V_2 = 20\ V$
3) $V_1 = 20\ V, V_2 = 10\ V$
4) $V_1 = 30\ V, V_2 = 15\ V$

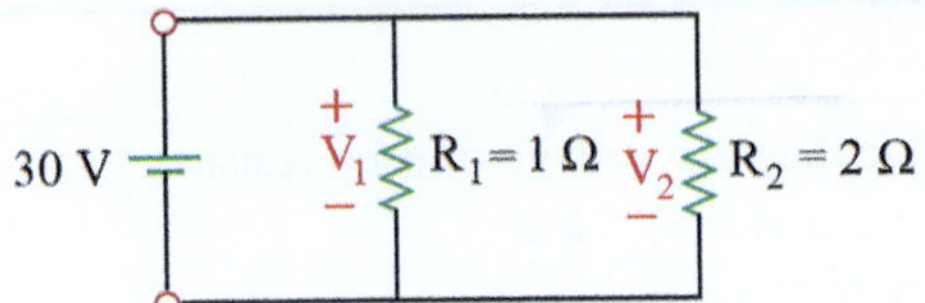

Fig. 3.23 The parallel connection of two resistors

Problem

3.35. In the circuit shown in Fig. 3.24, calculate the current of each resistor.

Difficulty level ● Easy ○ Normal ○ Hard

Calculation amount ● Small ○ Normal ○ Large

1) $I_1 = 30\,A, I_2 = 30\,A$
2) $I_1 = 30\,A, I_2 = 15\,A$
3) $I_1 = 10\,A, I_2 = 20\,A$
4) $I_1 = 20\,A, I_2 = 10\,A$

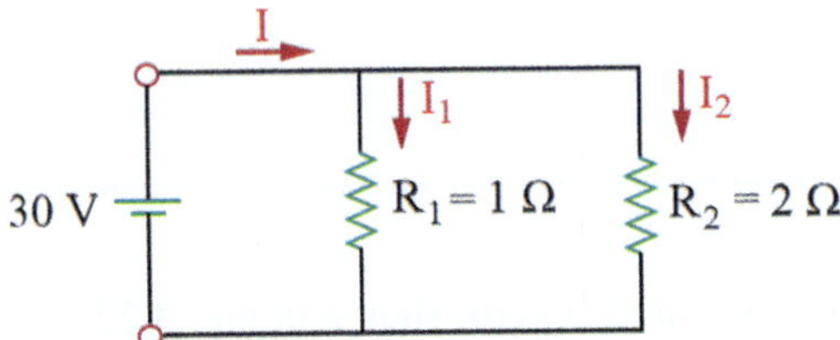

Fig. 3.24 The parallel connection of two resistors

Exercise

3.36. Calculate the current of the second resistor in the circuit shown in Fig. 3.25.

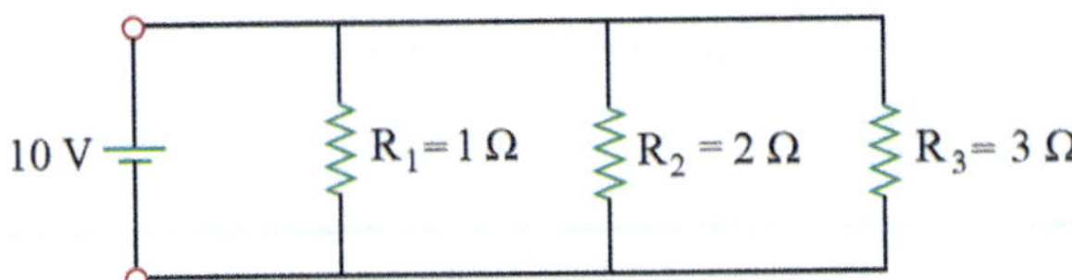

Fig. 3.25 The parallel connection of three resistors

Final Answer

$I_2 = 5\,A$

3.6 Electrical Power and Energy Wasted in Resistors

Problem

3.37. In the circuit shown in Fig. 3.26, calculate the power wasted by each resistor.

Difficulty level ○ Easy ● Normal ○ Hard

Calculation amount ○ Small ● Normal ○ Large

1) $P_1 = 100\ W, P_2 = 200\ W$
2) $P_1 = 10\ W, P_2 = 20\ W$
3) $P_1 = 20\ W, P_2 = 10\ W$
4) $P_1 = 200\ W, P_2 = 100\ W$

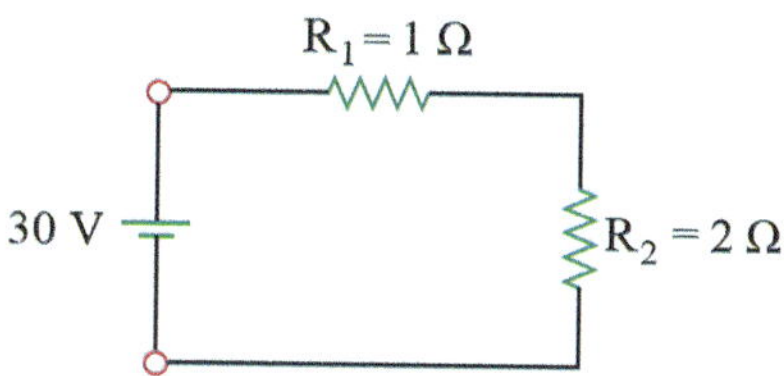

Fig. 3.26 The series connection of two resistors

Problem

3.38. In the circuit shown in Fig. 3.27, calculate the consumption power of each resistor.

Difficulty level ○ Easy ● Normal ○ Hard

Calculation amount ○ Small ● Normal ○ Large

1) $P_1 = 30\ W, P_2 = 15\ W$
2) $P_1 = 15\ W, P_2 = 30\ W$
3) $P_1 = 450\ W, P_2 = 900\ W$
4) $P_1 = 900\ W, P_2 = 450\ W$

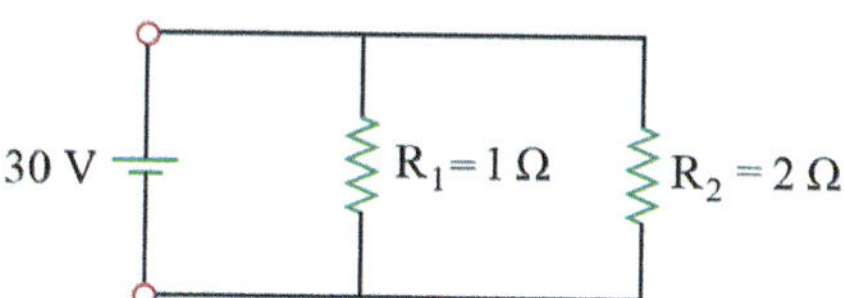

Fig. 3.27 The parallel connection of two resistors

Partially Solved Exercise

3.39. Calculate the power wasted by each resistor in the circuit shown in Fig. 3.28.

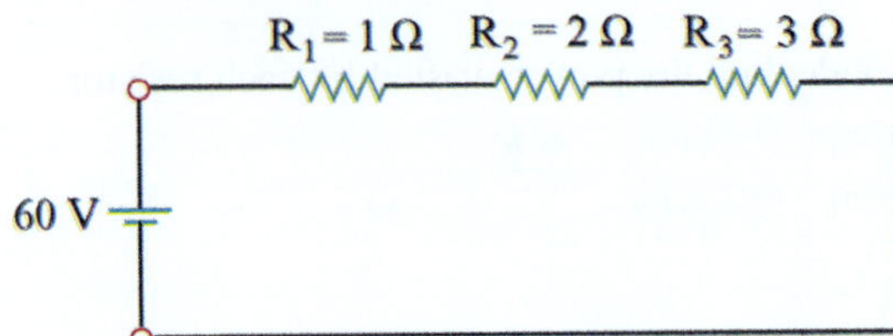

Fig. 3.28 The series connection of three resistors

Solution

The resistors are connected in series; therefore, their currents can be calculated as follows.

$$I_1 = I_2 = I_3 = \frac{60}{1+2+3} \Rightarrow I_1 = I_2 = I_3 = 10\ A$$

The power wasted or consumed by a resistor can be calculated by using one of the following relations.

$$P = RI^2 = VI = \frac{V^2}{R}$$

For this problem, the first relation is more appropriate. Hence:

$$P_1 = R_1 {I_1}^2 = 1 \times 10^2 \Rightarrow P_1 = 100\ W$$

$$P_2 = R_2 {I_2}^2 = 2 \times 10^2 \Rightarrow P_2 = 200\ W$$

$$P_3 = R_3 {I_3}^2 = 3 \times 10^2 \Rightarrow P_3 = 300\ W$$

Partially Solved Exercise

3.40. Calculate the power wasted by each resistor in the circuit shown in Fig. 3.29.

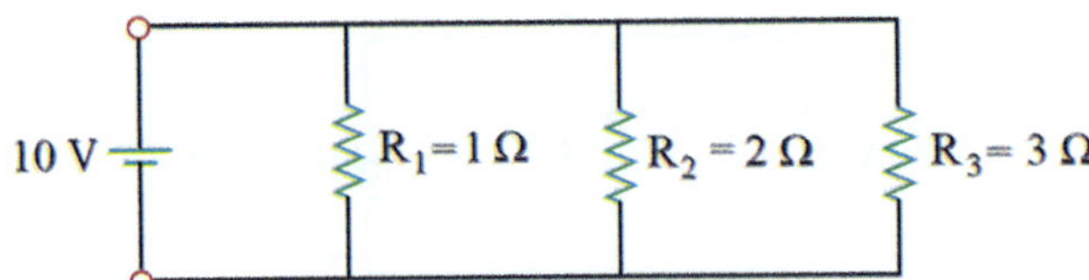

Fig. 3.29 The parallel connection of three resistors

Solution

The circuit is a parallel circuit; hence, the voltage of each resistor is equal to the voltage of the source. In other words:

$$V_1 = V_2 = V_3 = 10\ V$$

The power wasted or consumed by a resistor can be calculated by using one of the following relations.

$$P = RI^2 = VI = \frac{V^2}{R}$$

For this problem, the third relation is more appropriate. Hence:

$$P_1 = \frac{V_1^2}{R_1} = \frac{10^2}{1} \Rightarrow P_1 = 100\ W$$

$$P_2 = \frac{V_2^2}{R_2} = \frac{10^2}{2} \Rightarrow P_2 = 50\ W$$

$$P_3 = \frac{V_3^2}{R_3} = \frac{10^2}{3} \Rightarrow P_3 = 33.33\ W$$

References

1. Rahmani-Andebili, M., General Physics I - Practice Problems, Methods, and Solutions, Springer Nature, 2025.
2. Rahmani-Andebili, M., Calculus III – Practice Problems, Methods, and Solutions, Springer Nature, 2023.
3. Rahmani-Andebili, M., Calculus II – Practice Problems, Methods, and Solutions, Springer Nature, 2023.
4. Rahmani-Andebili, M., Calculus I (2nd Ed.) – Practice Problems, Methods, and Solutions, Springer Nature, 2023.
5. Rahmani-Andebili, M., Precalculus (2nd Ed.) – Practice Problems, Methods, and Solutions, Springer Nature, 2024.

4 Electrical Charge, Capacitance, Capacitor, Current, Resistance, and Resistor: Part B

Abstract

In this chapter, the problems of the third chapter are fully solved, in detail, step-by-step, and with different methods.

4.1 Capacitance and Equivalent Capacitance of a Circuit

4.1. Based on the information given in the problem, the initial capacitance of the flat capacitor is as follows [1–5].

$$C = \frac{\varepsilon_0 A}{d}$$

If part of the dielectric of a capacitor is replaced by a conductor with the surface area A and depth x, the capacitance of capacitor decreases and can be calculated as follows.

$$C' = \frac{\varepsilon_0 A}{d - x}$$

In this problem, $x = 0.5d$. therefore:

$$C' = \frac{\varepsilon_0 A}{d - 0.5d} = \frac{\varepsilon_0 A}{0.5d}$$

$$\Rightarrow C' = \frac{2\varepsilon_0 A}{d}$$

Choice (2) is the answer (Fig. 4.1).

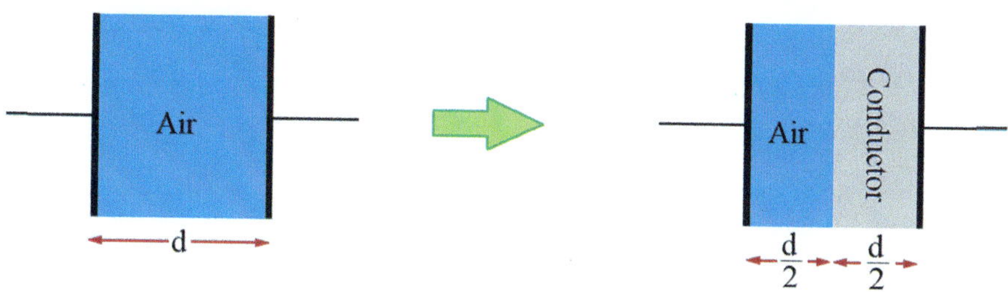

Fig. 4.1 A flat capacitor

M. Rahmani-Andebili, *General Physics II*, https://doi.org/10.1007/978-3-031-92866-6_4

4.2. The capacitor, shown in Fig. 4.2, includes two parts with the electrical permittivity ε_1 and ε_2. These two parts are, in fact, two individual capacitors with the surface area $\frac{A}{2}$ that have been connected in parallel. Thus, the total capacitance of the capacitors can be calculated as follows.

$$C = C_1 + C_2$$

$$\Rightarrow C = \frac{\varepsilon_1 \frac{A}{2}}{d} + \frac{\varepsilon_2 \frac{A}{2}}{d}$$

$$\Rightarrow C = \frac{A}{2d}(\varepsilon_1 + \varepsilon_2)$$

Choice (1) is the answer.

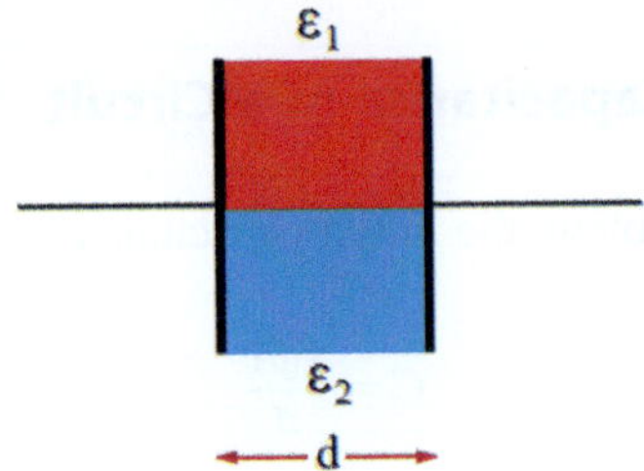

Fig. 4.2 A flat capacitor including two different electrical permittivity

Notes

In this problem, the relations below have been used.

The capacitance of a flat capacitor with the surface area A, distance d, and electrical permittivity ε is as follows.

$$C = \frac{\varepsilon A}{d} = \frac{\varepsilon_0 \varepsilon_r A}{d}$$

Herein, ε_0 and ε_r are the free space electrical permittivity and relative electrical permittivity.

The equivalent capacitance of n parallel capacitors can be calculated as follows.

$$C_{eq} = C_1 + \ldots + C_n$$

If $C_1 = \ldots = C_n = C$, we have:

$$C_{eq} = nC$$

4.3. The capacitor, shown in Fig. 4.3, includes two parts with the electrical permittivity ε_1 and ε_2. These two parts are, in fact, two individual capacitors with the distance $\frac{d}{2}$ that have been connected in series. Hence, the total capacitance of the capacitors can be calculated as follows.

$$C = \frac{C_1 C_2}{C_1 + C_2}$$

$$\Rightarrow C = \frac{\left(\frac{\varepsilon_1 A}{\frac{d}{2}}\right)\left(\frac{\varepsilon_2 A}{\frac{d}{2}}\right)}{\left(\frac{\varepsilon_1 A}{\frac{d}{2}}\right) + \left(\frac{\varepsilon_2 A}{\frac{d}{2}}\right)} = \frac{\frac{2\varepsilon_1 A}{d}\frac{2\varepsilon_2 A}{d}}{\frac{2A}{d}(\varepsilon_1 + \varepsilon_2)}$$

$$\Rightarrow C = \frac{2A}{d}\frac{\varepsilon_1 \varepsilon_2}{(\varepsilon_1 + \varepsilon_2)}$$

Choice (4) is the answer.

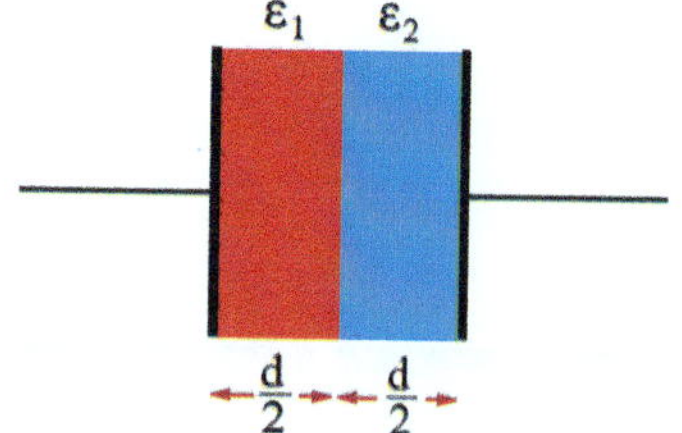

Fig. 4.3 A flat capacitor including two different electrical permittivity

Notes

In this problem, the relations below have been used.

The capacitance of a flat capacitor with the surface area A, distance d, and electrical permittivity ε is as follows.

$$C = \frac{\varepsilon A}{d} = \frac{\varepsilon_0 \varepsilon_r A}{d}$$

Herein, ε_0 and ε_r are the free space electrical permittivity and relative electrical permittivity.

The equivalent capacitance of n series capacitors can be calculated as follows.

$$\frac{1}{C_{eq}} = \frac{1}{C_1} + \ldots + \frac{1}{C_n}$$

If $C_1 = \ldots = C_n = C$, we have:

$$C_{eq} = \frac{C}{n}$$

If $n = 2$, we have:

$$\frac{1}{C_{eq}} = \frac{1}{C_1} + \frac{1}{C_2} \Rightarrow C_{eq} = \frac{C_1 C_2}{C_1 + C_2}$$

4.4. Based on the information given in the problem, we have:

$$r = 1 \ cm$$

$$d = 1 \ mm$$

$$\varepsilon_r = 1$$

$$\varepsilon_0 = 8.85 \times 10^{-12} \frac{C^2}{N.m^2}$$

As we know, the capacitance of a flat capacitor with the surface area A, distance d, and electrical permittivity ε is as follows.

$$C = \frac{\varepsilon A}{d} = \frac{\varepsilon_0 \varepsilon_r A}{d}$$

Herein, ε_0 and ε_r are the free space electrical permittivity and relative electrical permittivity.

Therefore, for this problem we can write:

$$C = \frac{\varepsilon_0 \pi r^2}{d}$$

$$\Rightarrow C = 8.85 \times 10^{-12} \times \frac{\pi(0.01)^2}{0.001}$$

$$\Rightarrow C = 2.78\ pF$$

Choice (4) is the answer.

4.5. Based on the information given in the problem, we have:

$$V = 120\ V$$

$$C = 2.78\ pF$$

The amount of charge stored in the capacitor can be calculated as follows.

$$q = VC$$

$$\Rightarrow q = 120 \times 2.78 \times 10^{-12}$$

$$\Rightarrow q = 333.6 \times 10^{-12}\ C$$

Choice (2) is the answer.

4.6. The circuit shown in Fig. 4.4 is the series connection of three capacitors. Therefore, the equivalent capacitance of the circuit can be calculated as follows.

$$\frac{1}{C_{eq}} = \frac{1}{C_1} + \frac{1}{C_2} + \frac{1}{C_3}$$

$$\Rightarrow \frac{1}{C_{eq}} = \frac{1}{2} + \frac{1}{6} + \frac{1}{3} = \frac{3+1+2}{6} = \frac{6}{6}$$

$$\Rightarrow C_{eq} = 1\ F$$

Choice (2) is the answer.

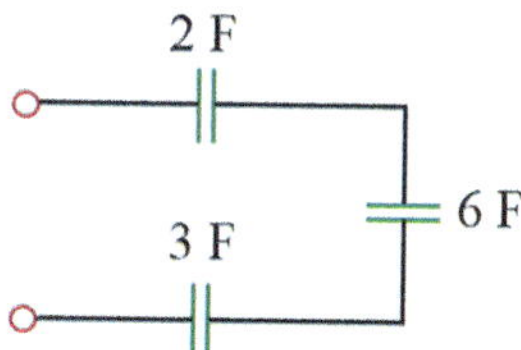

Fig. 4.4 The series connection of capacitors

Notes

In this problem, the relations below have been used.

The equivalent capacitance of n series capacitors can be calculated as follows.

$$\frac{1}{C_{eq}} = \frac{1}{C_1} + \ldots + \frac{1}{C_n}$$

If $C_1 = \ldots = C_n = C$, we have:

$$C_{eq} = \frac{C}{n}$$

If $n = 2$, we have:

$$\frac{1}{C_{eq}} = \frac{1}{C_1} + \frac{1}{C_2} \Rightarrow C_{eq} = \frac{C_1 C_2}{C_1 + C_2}$$

4.7. The circuit shown in Fig. 4.5 is the parallel connection of three capacitors. Therefore, the equivalent capacitance of the circuit can be calculated as follows.

$$C_{eq} = C_1 + C_2 + C_3$$

$$\Rightarrow C_{eq} = 2 + 3 + 6$$

$$\Rightarrow C_{eq} = 11\ F$$

Choice (1) is the answer.

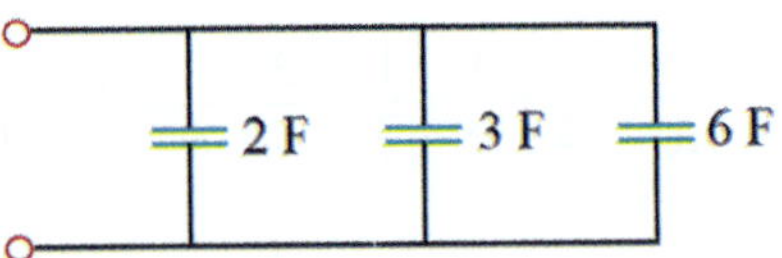

Fig. 4.5 The parallel connection of capacitors

Notes

In this problem, the relations below have been used.

The equivalent capacitance of n parallel capacitors can be calculated as follows.

$$C_{eq} = C_1 + \ldots + C_n$$

If $C_1 = \ldots = C_n = C$, we have:

$$C_{eq} = nC$$

4.8. The circuit shown in Fig. 4.6 is the series-parallel connection of three capacitors. Therefore, the equivalent capacitance of the circuit can be calculated as follows.

$$\frac{1}{C_{eq}} = \frac{1}{C_1} + \frac{1}{C_2 + C_3}$$

$$\Rightarrow \frac{1}{C_{eq}} = \frac{1}{2} + \frac{1}{1+1} = \frac{1}{2} + \frac{1}{2} = 1$$

$$\Rightarrow C_{eq} = 1\ F$$

Choice (4) is the answer.

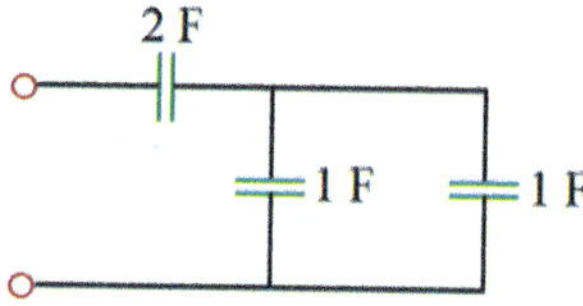

Fig. 4.6 The series-parallel connection of capacitors

Notes

In this problem, the relations below have been used.

The equivalent capacitance of n parallel capacitors can be calculated as follows.

$$C_{eq} = C_1 + \ldots + C_n$$

If $C_1 = \ldots = C_n = C$, we have:

$$C_{eq} = nC$$

The equivalent capacitance of n series capacitors can be calculated as follows.

$$\frac{1}{C_{eq}} = \frac{1}{C_1} + \ldots + \frac{1}{C_n}$$

If $C_1 = \ldots = C_n = C$, we have:

$$C_{eq} = \frac{C}{n}$$

If $n = 2$, we have:

$$\frac{1}{C_{eq}} = \frac{1}{C_1} + \frac{1}{C_2} \Rightarrow C_{eq} = \frac{C_1 C_2}{C_1 + C_2}$$

4.2 Electrical Charge and Voltage of Capacitors

4.10. In the first test, the capacitors are connected in series. Thus, the total capacitance of the circuit can be calculated as follows.

$$C_{tot,1} = \frac{C}{3}$$

In addition, in the second test, the capacitors are connected in parallel. Hence, the total capacitance of the circuit can be calculated as follows.

$$C_{tot,2} = 3C$$

On the other hand, the amount of charge stored in the capacitors can be calculated as follows.

$$q = VC$$

Therefore, the ratio of total charge of the capacitors in the first test to the second one is as follows.

$$\frac{q_1}{q_2} = \frac{VC_{tot,1}}{VC_{tot,2}} = \frac{\frac{C}{3}}{3C}$$

$$\Rightarrow \frac{q_1}{q_2} = \frac{1}{9}$$

Choice (1) is the answer.

Notes

In this problem, the relations below have been used.

The equivalent capacitance of n parallel capacitors can be calculated as follows.

$$C_{eq} = C_1 + \ldots + C_n$$

If $C_1 = \ldots = C_n = C$, we have:

$$C_{eq} = nC$$

The equivalent capacitance of n series capacitors can be calculated as follows.

$$\frac{1}{C_{eq}} = \frac{1}{C_1} + \ldots + \frac{1}{C_n}$$

If $C_1 = \ldots = C_n = C$, we have:

$$C_{eq} = \frac{C}{n}$$

If $n = 2$, we have:

$$\frac{1}{C_{eq}} = \frac{1}{C_1} + \frac{1}{C_2} \Rightarrow C_{eq} = \frac{C_1 C_2}{C_1 + C_2}$$

4.11. Choice (1) is correct because it shows the correct relationship between the charge, voltage, and capacitance of each capacitor as well as for the equivalent capacitor.

Choice (2) is correct because in a series circuit, the charge of each capacitor is equal, and they are equal to the total charge of the circuit.

Choice (3) is correct because in a series circuit, the voltage of each capacitor is inversely proportional to its size. In other words, the more capacitance, the less voltage.

Choice (4) is wrong based on the rule mentioned in Choice (3).

Choice (4) is the answer (Fig. 4.7).

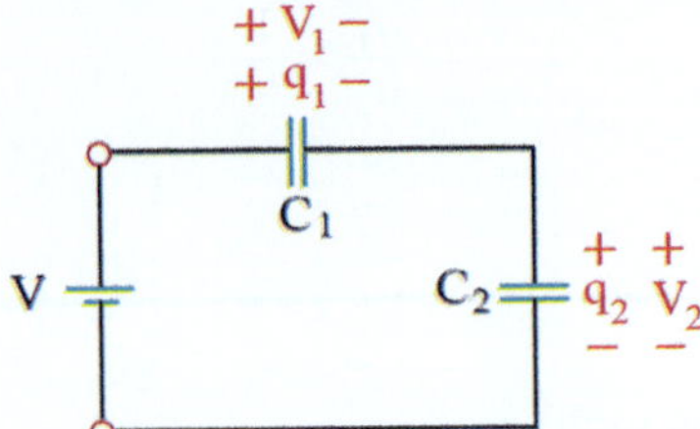

Fig. 4.7 The series connection of two capacitors

4.12. Choice (1) is correct because it shows the correct relationship between the charge, voltage, and capacitance of each capacitor as well as for the equivalent capacitor.

Choice (2) is correct because in a parallel circuit, the voltage of each capacitor is equal, and they are equal to the voltage of the source.

Choice (3) is correct because in a parallel circuit, the charge of each capacitor is proportional to its size. In other words, the more capacitance, the more charge.

Choice (4) is wrong based on the rule mentioned in Choice (3).

Choice (4) is the answer (Fig. 4.8).

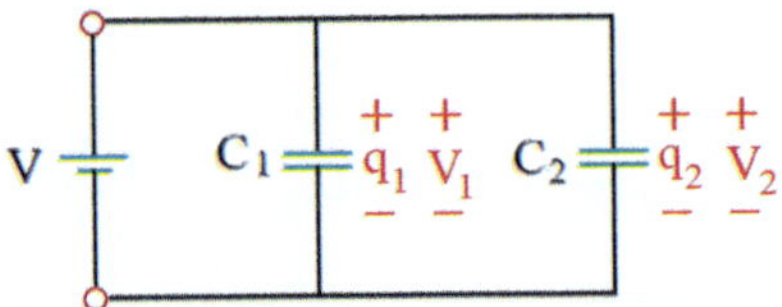

Fig. 4.8 The parallel connection of two capacitors

4.13. Based on the information given in the problem, we have:

$$C_1 = 2\ \mu F$$

$$C_2 = 3\ \mu F$$

$$V = 150\ V$$

The voltage across the C_1 can be calculated as follows.

$$V_1 = \frac{C_2}{C_1 + C_2} V$$

Therefore:

$$V_1 = \frac{2}{2+3} \times 150$$

$$\Rightarrow V_1 = 60\ V$$

Choice (2) is the answer (Fig. 4.9).

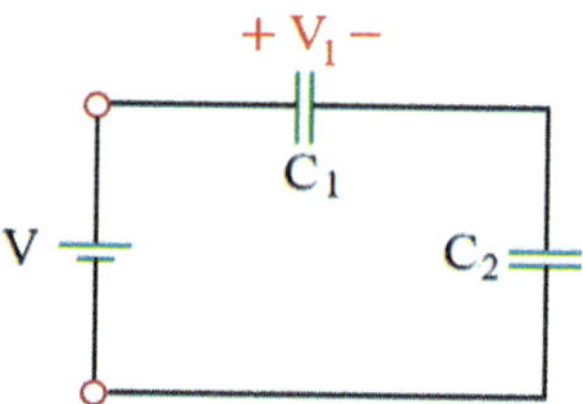

Fig. 4.9 The series connection of two capacitors

4.14. Based on the information given in the problem, we have:

$$C_1 = 4\ \mu F$$

$$C_2 = 6\ \mu F$$

$$V = 200\ V$$

As we know, the voltage of C_1 can be calculated as follows.

$$V_1 = \frac{C_2}{C_1 + C_2} V$$

$$\Rightarrow V_1 = \frac{6}{4+6} \times 200 = 120\ V$$

Moreover, the charge of a capacitor can be calculated as follows.

$$q = VC$$

Thus:

$$q_1 = V_1 C_1 \Rightarrow q_1 = 120 \times 4 \times 10^{-6}$$

$$\Rightarrow q_1 = 4.8 \times 10^{-4}\ C$$

Choice (1) is the answer (Fig. 4.10).

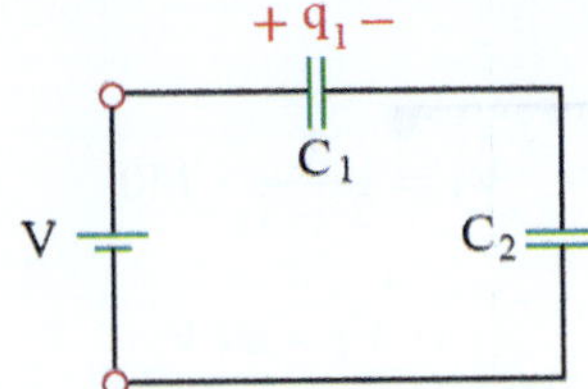

Fig. 4.10 The series connection of two capacitors

4.15. Based on the information given in the problem, we have:

$$C_1 = 2\ \mu F$$

$$C_2 = 3\ \mu F$$

$$V = 150\ V$$

The voltage across C_1 can be calculated as follows.

$$V_1 = V$$

Therefore:

$$\Rightarrow V_1 = 150\ V$$

Choice (4) is the answer (Fig. 4.11).

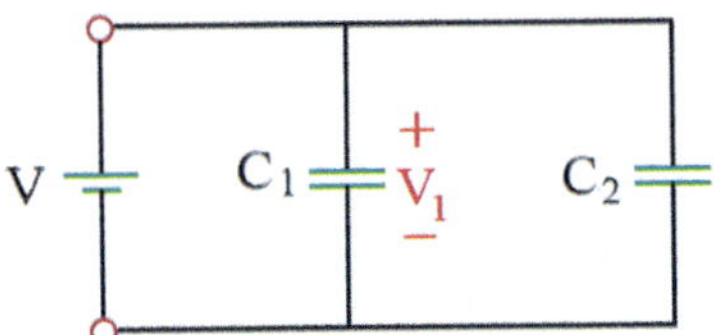

Fig. 4.11 The parallel connection of two capacitors

4.16. Based on the information given in the problem, we have:

$$C_1 = 4\ \mu F$$

$$C_2 = 6\ \mu F$$

$$V = 200\ V$$

As we know, the voltage of C_1 is equal to the voltage of the voltage source. In other words:

$$V_1 = 200\ V$$

Moreover, the charge of a capacitor can be calculated as follows.

$$q = VC$$

Thus:

$$q_1 = V_1 C_1 \Rightarrow q_1 = 200 \times 4 \times 10^{-6}$$

$$\Rightarrow q_1 = 8 \times 10^{-4}\ C$$

Choice (3) is the answer (Fig. 4.12).

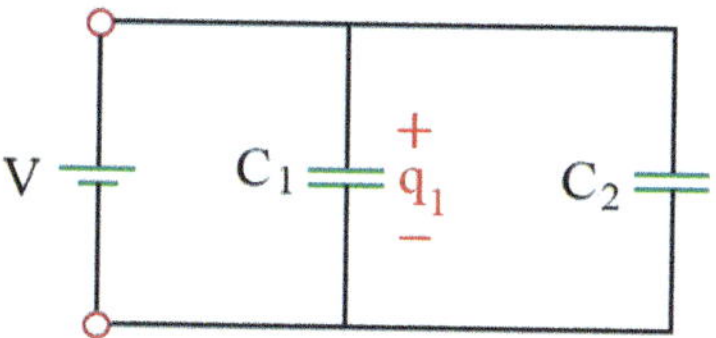

Fig. 4.12 The parallel connection of two capacitors

4.17. This problem is interesting. Based on the information given in the problem, we have:

$$C_1 = 2\ \mu F$$

$$V_1 = 100\ V$$

$$C_2 = 3\ \mu F$$

$$q_2 = 0\ C$$

When two capacitors are connected to each other, their voltage, after the connection, can be calculated using the relation below.

$$V = \frac{q_1 \pm q_2}{C_{eq}} = \frac{V_1 C_1 \pm V_2 C_2}{C_1 + C_2}$$

In the previous relation, the addition operator is applied if the similar polarities of the capacitors are connected to each other; otherwise, the subtraction operator is used.

Therefore:

$$V = \frac{2 \times 10^{-6} \times 100 \pm 0}{2 \times 10^{-6} + 3 \times 10^{-6}} = \frac{2 \times 10^{-4}}{5 \times 10^{-6}} = 40\ V$$

Since the capacitors are in parallel, we have:

$$V_1 = V_2 = 40\ V$$

Choice (3) is the answer.

4.18. This problem is interesting. Since the capacitor is connected to the voltage source, its voltage remains constant and equal to the voltage of the voltage source.

By increasing the distance between the plates of the capacitor, the capacitance of the capacitor decreases as can be seen in the following.

$$\downarrow C = \frac{\varepsilon_0 A}{\uparrow d}$$

However, based on the relation below, the charge of the capacitor decreases.

$$\downarrow q = VC \downarrow$$

Therefore, by increasing the distance between the plates of a capacitor connected to a battery, the voltage of the capacitor remains constant but its capacity and charge decrease. Choice (4) is the answer.

4.19. Like the previous problem, this one is interesting. Since the capacitor is disconnected from the voltage source, its charge remains constant even if its capacitance changes.

By increasing the distance between the plates of the capacitor, the capacitance of the capacitor decreases as can be seen in the following.

$$\downarrow C = \frac{\varepsilon_0 A}{\uparrow d}$$

Nonetheless, based on the following relation, the voltage of the capacitor increases.

$$\uparrow V = \frac{q}{\downarrow C}$$

Therefore, by increasing the distance between the plates of a capacitor disconnected from its battery, the charge of the capacitor remains constant, its capacity decreases, and its voltage increases. Choice (3) is the answer.

4.3 Electrical Energy Stored in Capacitors

4.20. Based on the information given in the problem, we have:

$$C_1 = 1\ \mu F$$

$$C_2 = 2\ \mu F$$

$$C_3 = 4\ \mu F$$

$$q_1 = 100\ \mu C$$

Since the capacitors have been connected in parallel, their voltages are equal. Thus:

$$V_1 = V_2 = V_3 = \frac{q_1}{C_1} = \frac{100 \times 10^{-6}}{10^{-6}} = 100\ V$$

The electrical energy stored in C_3 can be calculated as follows.

$$W_3 = \frac{1}{2} C_3 V_3^2$$

$$\Rightarrow W_3 = 4 \times 10^{-6} \times 100^2$$

$$\Rightarrow W_3 = 0.04\ J = 40\ mJ$$

Choice (1) is the answer.

4.4 Resistance and Equivalent Resistance of a Circuit

4.23. Based on the information given in the problem, we have:

$$2r = 1\ mm$$

$$l = 2\ m$$

$$R = 0.1\ \Omega$$

As we know, the resistance of a wire can be calculated as follows.

$$R = \rho \frac{l}{A}$$

Herein, ρ, l, and A, are the special resistance (resistivity), length, and cross-sectional area of the wire, respectively.

Thus, for this problem, we can write:

$$R = \rho \frac{l}{\pi r^2}$$

$$\Rightarrow 0.1 = \rho \frac{2}{\pi (0.5 \times 10^{-3})^2}$$

$$\Rightarrow \rho = \frac{0.1 \times 0.25\pi \times 10^{-6}}{2}$$

$$\Rightarrow \rho = 3.92 \times 10^{-8}\ \Omega m$$

Choice (1) is the answer.

4.24. Based on the information given in the problem, we have:

$$l_1 = 2l_2$$

$$d_1 = 0.5d_2 \Rightarrow r_1 = 0.5r_2$$

$$\rho_1 = \frac{1}{3}\rho_2$$

As we know, the resistance of a wire can be calculated as follows.

$$R = \rho \frac{l}{A}$$

Herein, ρ, l, and A, are the special resistance (resistivity), length, and cross-sectional area of the wire, respectively.

Thus, for this problem, we can write:

$$R = \rho \frac{l}{\pi r^2}$$

$$\Rightarrow \frac{R_1}{R_2} = \frac{\rho_1}{\rho_2} \times \frac{l_1}{l_2} \times \left(\frac{r_2}{r_1}\right)^2$$

$$\Rightarrow \frac{R_1}{R_2} = \frac{1}{3} \times 2 \times (2)^2$$

$$\Rightarrow \frac{R_1}{R_2} = \frac{8}{3}$$

Choice (4) is the answer.

4.25. The total resistance of the circuit can be calculated as follows.

$$R_{eq} = 2 + 6 + 3$$

$$\Rightarrow R_{eq} = 11\ \Omega$$

Choice (2) is the answer (Fig. 4.13).

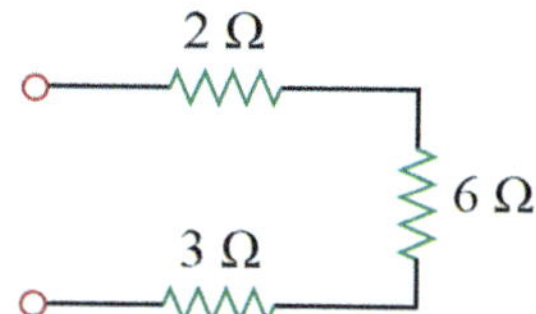

Fig. 4.13 The series connection of resistors

Notes

In this problem, the relations below have been used.

The equivalent resistance of n series resistors can be calculated as follows.

$$R_{eq} = R_1 + \ldots + R_n$$

If $R_1 = \ldots = R_n = R$, we have:

$$R_{eq} = nR$$

4.26. The total resistance of the circuit can be calculated as follows.

$$\frac{1}{R_{eq}} = \frac{1}{2} + \frac{1}{3} + \frac{1}{6} = \frac{3+2+1}{6} = \frac{6}{6}$$

$$\Rightarrow R_{eq} = 1\ \Omega$$

Choice (1) is the answer (Fig. 4.14).

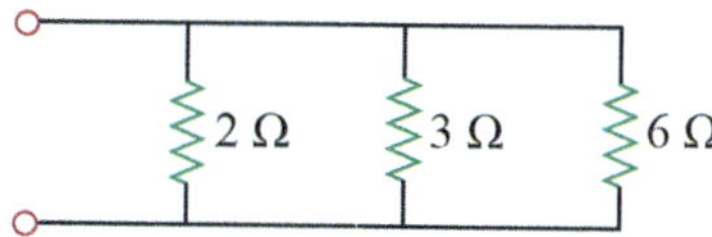

Fig. 4.14 The parallel connection of resistors

Notes

In this problem, the relations below have been used.

The equivalent resistance of n parallel resistors can be calculated as follows.

$$\frac{1}{R_{eq}}=\frac{1}{R_1}+\ldots+\frac{1}{R_n}$$

If $R_1=\ldots=R_n=R$, we have:

$$R_{eq}=\frac{R}{n}$$

If $n=2$, we have:

$$\frac{1}{R_{eq}}=\frac{1}{R_1}+\frac{1}{R_2}\Rightarrow R_{eq}=\frac{R_1R_2}{R_1+R_2}$$

4.27. In the circuit, the resistors are connected in series-parallel from. Therefore:

$$R_{eq}=2+1||1$$

$$\Rightarrow R_{eq}=2+\frac{1\times 1}{1+1}$$

$$\Rightarrow R_{eq}=2.5\ \Omega$$

Choice (4) is the answer (Fig. 4.15).

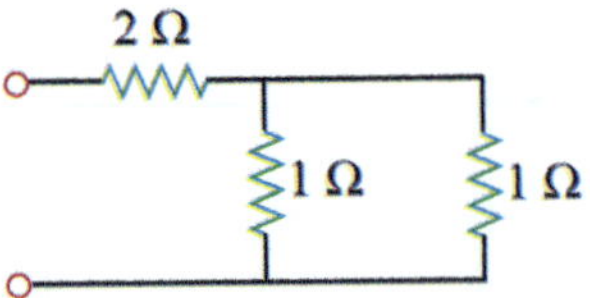

Fig. 4.15 The series-parallel connection of resistors

Notes

In this problem, the relations below have been used.

The equivalent resistance of n series resistors can be calculated as follows.

$$R_{eq} = R_1 + \ldots + R_n$$

If $R_1 = \ldots = R_n = R$, we have:

$$R_{eq} = nR$$

The equivalent resistance of n parallel resistors can be calculated as follows.

$$\frac{1}{R_{eq}} = \frac{1}{R_1} + \ldots + \frac{1}{R_n}$$

If $R_1 = \ldots = R_n = R$, we have:

$$R_{eq} = \frac{R}{n}$$

If $n = 2$, we have:

$$\frac{1}{R_{eq}} = \frac{1}{R_1} + \frac{1}{R_2} \Rightarrow R_{eq} = \frac{R_1 R_2}{R_1 + R_2}$$

4.5 Ohm's Law and Electrical Current and Voltage of Resistors

4.29. Choice (1) is correct because it shows the correct relationship between the current, voltage, and resistance of each resistor as well as for the equivalent resistor.

Choice (2) is correct because in a series circuit, the current of each resistor is equal, and they are equal to the total current of the circuit.

Choice (3) is correct because in a series circuit, the voltage of each resistor is proportional to its size. In other words, the more resistance, the more voltage.

Choice (4) is wrong based on the rule mentioned in Choice (3).

Choice (4) is the answer (Fig. 4.16).

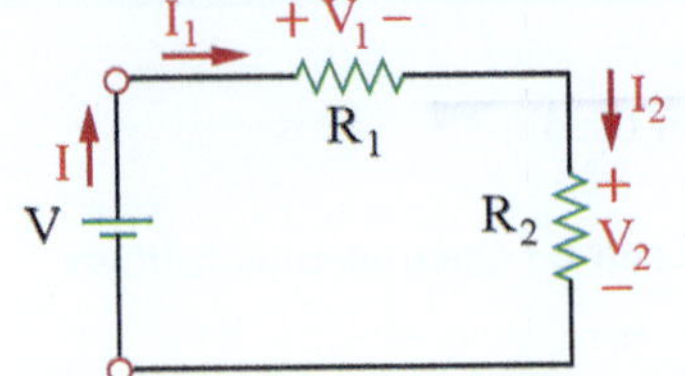

Fig. 4.16 The series connection of two resistors

4.30. Choice (1) is correct because it shows the correct relationship between the current, voltage, and resistance of each resistor as well as for the equivalent resistor.

Choice (2) is correct because in a parallel circuit, the voltage of each resistor is equal, and they are equal to the voltage of the source.

Choice (3) is correct because in a parallel circuit, the current of each resistor is inversely proportional to its size. In other words, the more resistance, the less current.

Choice (4) is wrong based on the rule mentioned in Choice (3).

Choice (4) is the answer (Fig. 4.17).

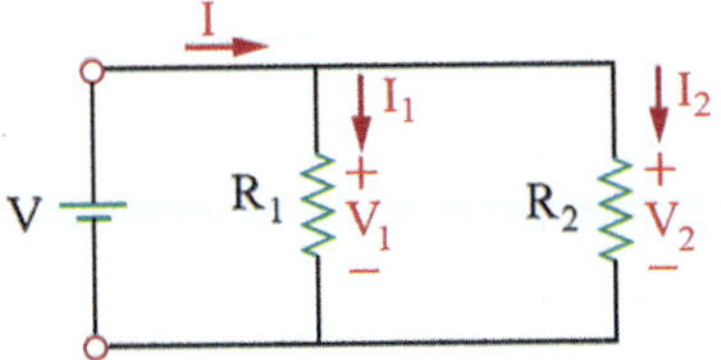

Fig. 4.17 The parallel connection of two resistors

4.31. In the circuit illustrated in Fig. 4.18, the resistors are connected in series. Therefore, their currents are equal and can be calculated as follows.

$$\Rightarrow I_1 = I_2 = I = \frac{30}{1+2}$$

$$\Rightarrow I_1 = I_2 = 10\,A$$

Choice (4) is the answer.

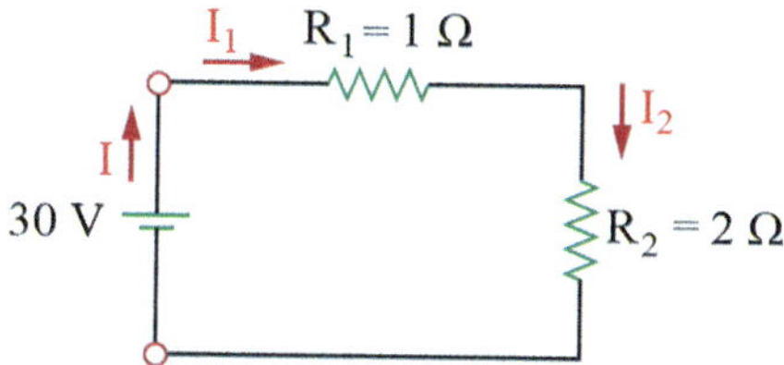

Fig. 4.18 The series connection of two resistors

4.32. The circuit is a series circuit; therefore, the voltage of each resistor can be calculated as follows.

$$V_1 = \frac{R_1}{R_1 + R_2} V$$

$$V_2 = \frac{R_2}{R_1 + R_2} V$$

Hence:

$$V_1 = \frac{1}{1+2} \times 30 \Rightarrow V_1 = 10\ V$$

$$V_2 = \frac{2}{1+2} \times 30 \Rightarrow V_2 = 20\ V$$

Choice (3) is the answer (Fig. 4.19).

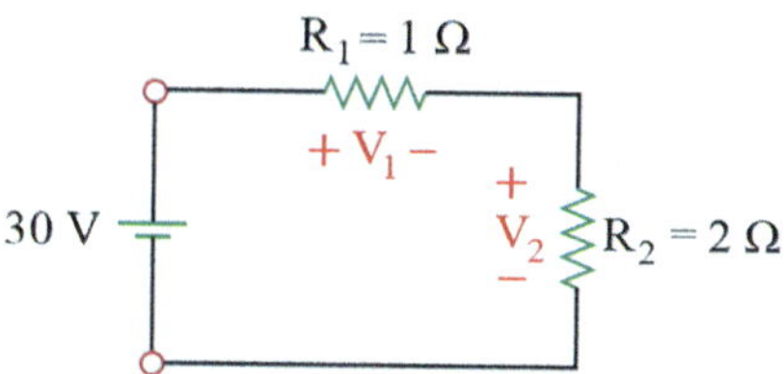

Fig. 4.19 The series connection of two resistors

4.34. The circuit is a parallel circuit; therefore, the voltage of each resistor is equal to the voltage of the source. Therefore:

$$V_1 = V_2 = 30\ V$$

Choice (1) is the answer (Fig. 4.20).

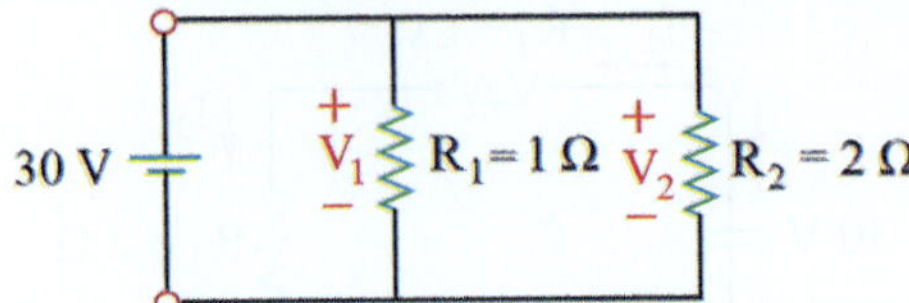

Fig. 4.20 The parallel connection of two resistors

4.35. The circuit is a parallel circuit; therefore, the current of resistors can be calculated as follows.

$$I_1 = \frac{V_1}{R_1} = \frac{V}{R_1}$$

$$I_2 = \frac{V_2}{R_2} = \frac{V}{R_2}$$

Therefore:

$$I_1 = \frac{30}{1} \Rightarrow I_1 = 30\ A$$

$$I_1 = \frac{30}{2} \Rightarrow I_1 = 15\ A$$

Choice (2) is the answer (Fig. 4.21).

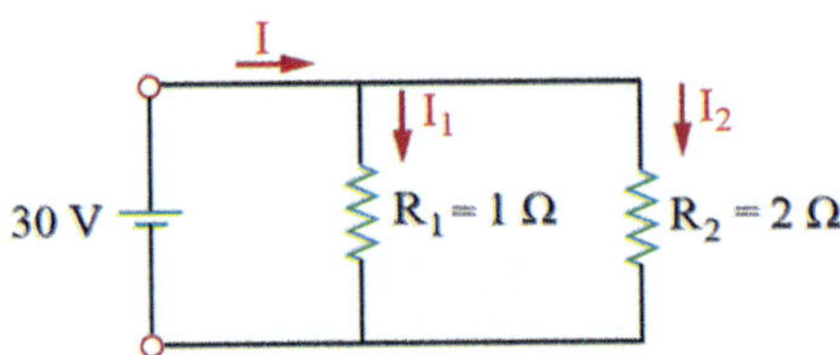

Fig. 4.21 The parallel connection of two resistors

4.6 Electrical Power and Energy Wasted in Resistors

4.37. In the circuit illustrated in Fig. 4.22, the resistors are connected in series. Therefore, their currents can be calculated as follows.

$$I_1 = I_2 = \frac{30}{1+2} \Rightarrow I_1 = I_2 = 10\ A$$

The power wasted or consumed by a resistor can be calculated by using one of the following relations.

$$P = RI^2$$

$$P = VI$$

$$P = \frac{V^2}{R}$$

For this problem, the first relation is more appropriate. Hence:

$$P_1 = R_1 {I_1}^2 = 1 \times 10^2 \Rightarrow P_1 = 100\ W$$

$$P_2 = R_2 {I_2}^2 = 2 \times 10^2 \Rightarrow P_2 = 200\ W$$

Choice (1) is the answer.

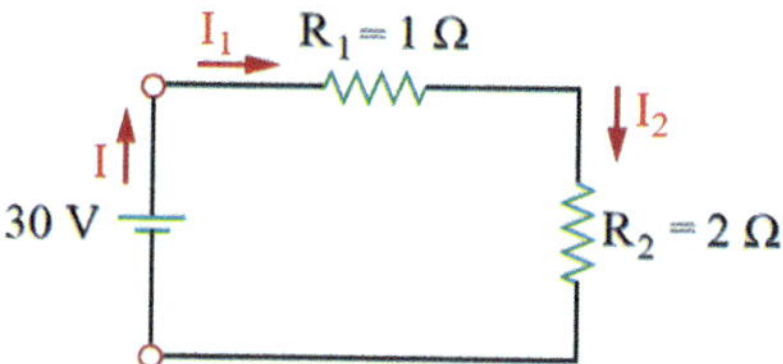

Fig. 4.22 The series connection of two resistors

4.38. The circuit is a parallel circuit; therefore, the voltage of each resistor is equal to the voltage of the source. In other words:

$$V_1 = V_2 = 30\ V$$

The power wasted or consumed by a resistor can be calculated by using one of the following relations.

$$P = RI^2$$

$$P = VI$$

$$P = \frac{V^2}{R}$$

For this problem, the third relation is more appropriate. Hence:

$$P_1 = \frac{{V_1}^2}{R_1} = \frac{30^2}{1} \Rightarrow P_1 = 900\ W$$

$$P_2 = \frac{{V_2}^2}{R_2} = \frac{30^2}{2} \Rightarrow P_2 = 450\ W$$

Choice (4) is the answer (Fig. 4.23).

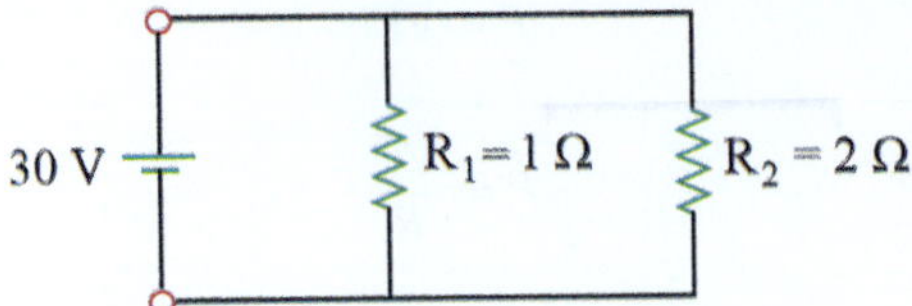

Fig. 4.23 The parallel connection of two resistors

References

1. Rahmani-Andebili, M., General Physics I - Practice Problems, Methods, and Solutions, Springer Nature, 2025.
2. Rahmani-Andebili, M., Calculus III – Practice Problems, Methods, and Solutions, Springer Nature, 2023.
3. Rahmani-Andebili, M., Calculus II – Practice Problems, Methods, and Solutions, Springer Nature, 2023.
4. Rahmani-Andebili, M., Calculus I (2nd Ed.) – Practice Problems, Methods, and Solutions, Springer Nature, 2023.
5. Rahmani-Andebili, M., Precalculus (2nd Ed.) – Practice Problems, Methods, and Solutions, Springer Nature, 2024.

Magnetic Field: Part A

5

Abstract

In this chapter, the basic and advanced problems of Magnetic Field are studied. The subjects include Magnetic Field Intensity Around a Long Current-Carrying Wire, Magnetic Field Intensity at the Center of a Circle and a Circular Arc, Magnetic Energy, Magnetic Energy Density, and Magnetic Force. Herein, different types of problems and exercises are presented that are categorized as follows.

- ***Problems with detailed solution***: They have been designed to teach students the subjects in detail. Moreover, they have been categorized in different levels based on their difficulty levels (easy, normal, and hard) and calculation amounts (small, normal, and large).
- ***Partially solved exercises***: They have been designed to encourage students to practice problems while guiding them through the problem-solving procedure and hinting the required formulas.
- ***Exercises with final answer***: They have been designed to encourage students to practice more by themselves while hinting them by the final answer as well as to help instructors to give tests or quizzes.

5.1 Magnetic Field Intensity Around a Long Current-Carrying Wire

Problem

5.1. The magnitude of magnetic flux density at one meter distance from a long current-carrying wire is about $4 \times 10^{-7}\,T$. Calculate the amount of current of the wire placed in a free space. Herein, assume that $\mu_0 = 4\pi \times 10^{-7}$ [1–5].

Difficulty level ● Easy ○ Normal ○ Hard

Calculation amount ● Small ○ Normal ○ Large

1) $\frac{2}{\pi}A$
2) $0\,A$
3) $2\,A$
4) $1\,A$

Exercise

5.2. Calculate the magnetic field intensity in free space at a distance from a long current-carrying wire. Herein, assume that distance from the wire and the current of the wire are infinite.

Final Answer

$H = \frac{1}{2\pi}\,A/m$

M. Rahmani-Andebili, *General Physics II*, https://doi.org/10.1007/978-3-031-92866-6_5

Problem

5.3. As is shown in Fig. 5.1, two parallel and long wires carry the currents of $I_1 = 1\,A$ and $I_2 = 2\,A$ at the same direction. The distance between them is 0.2 m. Calculate the resultant magnetic field intensity at midpoint between them.

Difficulty level ○ Easy ● Normal ○ Hard

Calculation amount ○ Small ● Normal ○ Large

1) $\frac{10}{\pi}\,A/m$
2) $0\,A/m$
3) $\frac{15}{\pi}\,A/m$
4) $\frac{5}{\pi}\,A/m$

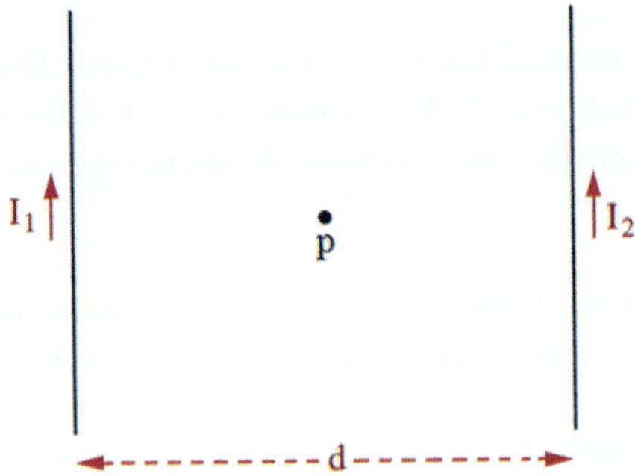

Fig. 5.1 Two parallel current-carrying wires

Partially Solved Exercise

5.4. Solve the previous problem by assuming that the wires carry the currents in the opposite directions, as is illustrated in Fig. 5.2.

Fig. 5.2 Two parallel current-carrying wires

Solution

Based on the information given in the problem, we have:

$$I_1 = 1\,A$$

$$I_2 = 2\,A$$

$$d = 0.2\,m$$

Figure 5.3 illustrates the magnetic field intensity of each wire at point p based on the right-hand rule. As can be seen, the field intensities are in the same direction (inward). Therefore:

$$H_{tot} = H_2 + H_1$$

Moreover, the magnitude of magnetic field intensity of each wire can be calculated as follows.

$$H_1 = \frac{I_1}{2\pi r_1} = \frac{(\quad)}{2\pi \times (\quad)} = (\quad)\,A/m$$

$$H_2 = \frac{I_2}{2\pi r_2} = \frac{(\quad)}{2\pi \times (\quad)} = (\quad)\,A/m$$

Thus:

$$H_{tot} = (\quad) + (\quad) \Rightarrow H_{tot} = \frac{15}{\pi}\,A/m$$

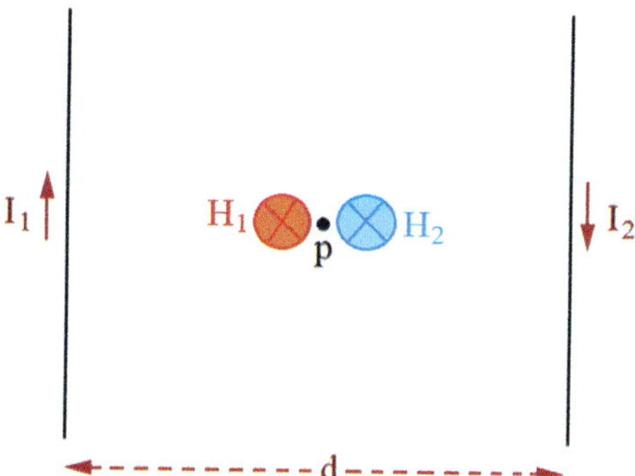

Fig. 5.3 Two parallel current-carrying wires

Partially Solved Exercise

5.5. As is shown in Fig. 5.4, two parallel and long wires carry the currents of $I_1 = I_2 = 2\pi\ A$ at the same direction. The distance between them is 1 m. Calculate the resultant magnetic field intensity at point p which is on the same plane but one meter away from the second wire.

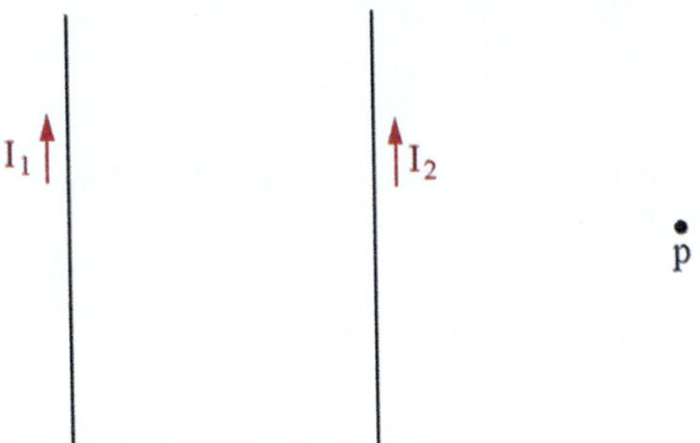

Fig. 5.4 Two parallel current-carrying wires

Solution

Based on the information given in the problem, we have:

$$I_1 = 1\ A$$

$$I_2 = 1\ A$$

$$r_1 = 2\ m$$

$$r_2 = 1\ m$$

Figure 5.5 illustrates the magnetic field intensity of each wire at point p based on the right-hand rule. As can be seen, the field intensities are in the same direction (inward). Therefore:

$$H_{tot} = H_2 + H_1$$

Moreover, the magnitude of magnetic field intensity of the wires can be calculated as follows.

$$H_1 = \frac{I_1}{2\pi r_1} = \frac{(\quad)}{2\pi \times (\quad)} = (\quad)\ A/m$$

$$H_2 = \frac{I_2}{2\pi r_2} = \frac{(\quad)}{2\pi \times (\quad)} = (\quad)\ A/m$$

Thus:

$$H_{tot} = (\quad) + (\quad) \Rightarrow H_{tot} = 1.5\ A/m$$

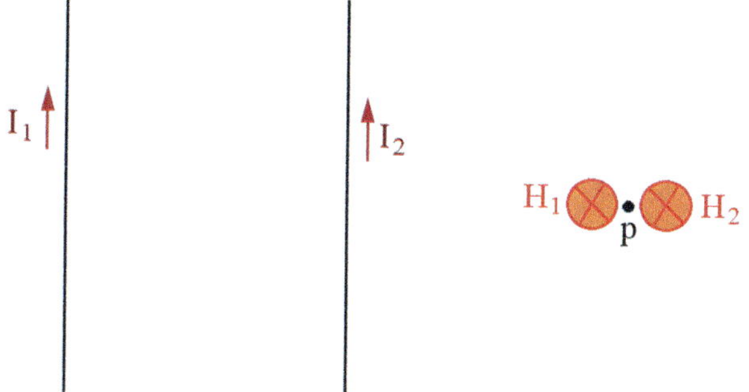

Fig. 5.5 Two parallel current-carrying wires

Exercise

5.6. Solve Problem 5.5 by assuming that the wires carry the currents in the opposite directions, as is illustrated in Fig. 5.6. Moreover, determine the direction of the resultant magnetic field intensity.

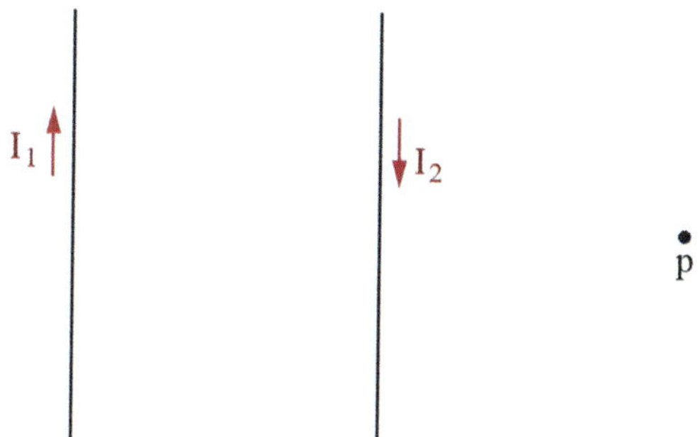

Fig. 5.6 Two parallel current-carrying wires

Final Answer

$H_{tot} = 0.5\frac{A}{m}, \odot$

Exercise

5.7. Two parallel and long wires carry different amounts of current but in the same direction. Determine the location of a point where the resultant magnetic field intensity is zero.

Final Answer

The point is between the wires and close to the wire with smaller current.

Exercise

5.8. Two parallel and long wires carry different amounts of current but in opposite directions. Determine the location of a point where the resultant magnetic field intensity is zero.

Final Answer

The point is outside of the wires and close to the wire with smaller current.

Problem

5.9. Figure 5.7 shows two parallel and long current-carrying wires, where the resultant magnetic field intensity at point p is zero. Calculate the distance between the wires if $I_1 = 4\ A$, $I_2 = 8\ A$, and $d_1 = 0.1\ m$.

Difficulty level ○ Easy ○ Normal ● Hard

Calculation amount ○ Small ● Normal ○ Large

1) 0.2 m
2) 0.3 m
3) 0.4 m
4) 0.5 m

Fig. 5.7 Two parallel current-carrying wires

Problem

5.10. The resultant magnetic field intensity at point p, illustrated in Fig. 5.8, is zero. Calculate the distance between the wires if $I_1 = 3\ A$, $I_2 = 1\ A$, and $d_2 = 0.1\ m$.

Difficulty level ○ Easy ○ Normal ● Hard

Calculation amount ○ Small ● Normal ○ Large

1) 0.2 m
2) 0.3 m
3) 0.4 m
4) 0.5 m

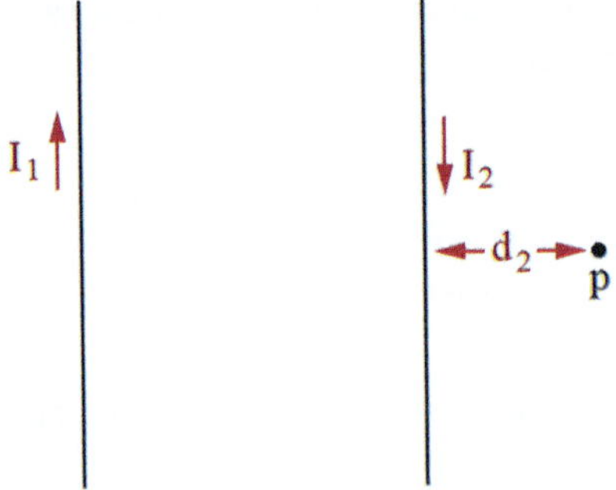

Fig. 5.8 Two parallel current-carrying wires

Problem

5.11. In the system shown in Fig. 5.9, the two parallel wires carry the currents $I_1 = 3\,A$ and $I_2 = 4\,A$ in opposite directions. Calculate the magnitude of resultant magnetic field intensity at point p. Herein, $d = 2\ m$.

Difficulty level ○ Easy ○ Normal ● Hard 

Calculation amount ○ Small ● Normal ○ Large

1) $H = \frac{5}{2\sqrt{2}\pi}\ A/m$
2) $H = \frac{5}{2\sqrt{2}}\ A/m$
3) $H = \frac{1}{2\sqrt{2}\pi}\ A/m$
4) $H = \frac{5}{\sqrt{2}\pi}\ A/m$

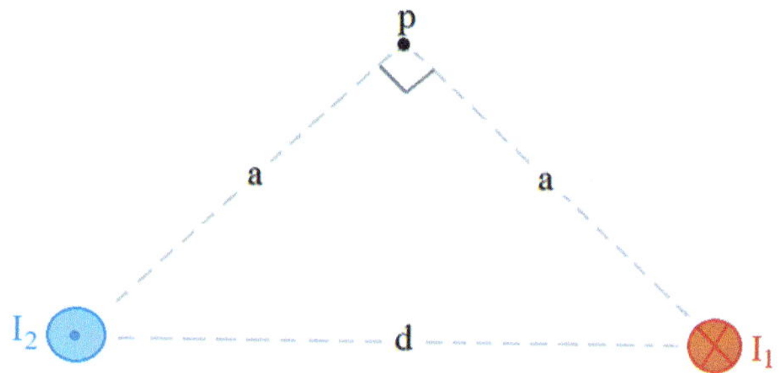

Fig. 5.9 Two parallel current-carrying wires

5.2 Magnetic Field Intensity at the Center of a Circle and a Circular Arc

Problem

5.12. Calculate the magnetic flux density in free space at the center of a circle with the radius of 0.6 m carrying 10 A current. Herein, assume that $\pi \approx 3$ and $\mu_0 = 4\pi \times 10^{-7}$.

Difficulty level ● Easy ○ Normal ○ Hard 

Calculation amount ● Small ○ Normal ○ Large

1) $\frac{10^{-5}}{\pi}\ T$
2) $10^{-7}\ T$
3) $10^{-6}\ T$
4) $10^{-5}\ T$

Exercise

5.13. Calculate the magnetic field intensity in free space at the center of a circle carrying a current. Herein, assume that radius and current of the circle are infinite.

Final Answer

$H = 0.5\ A/m$

Problem

5.14. Calculate the magnitude of magnetic field intensity at the center of the circular arc shown in Fig. 5.10.

Difficulty level ○ Easy ● Normal ○ Hard

Calculation amount ● Small ○ Normal ○ Large

1) $\frac{I\theta}{2\pi R}$
2) $\frac{I}{2R}$
3) $\frac{I\theta}{4R}$
4) $\frac{I\theta}{4\pi R}$

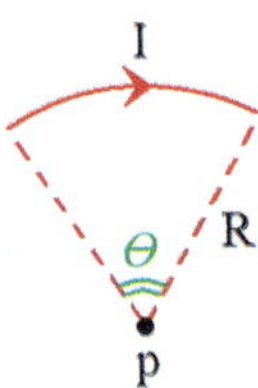

Fig. 5.10 Magnetic field intensity at the center of a circular arc

Partially Solved Exercise

5.15. Calculate the magnitude of magnetic field intensity at point p in the circuit shown in Fig. 5.11.

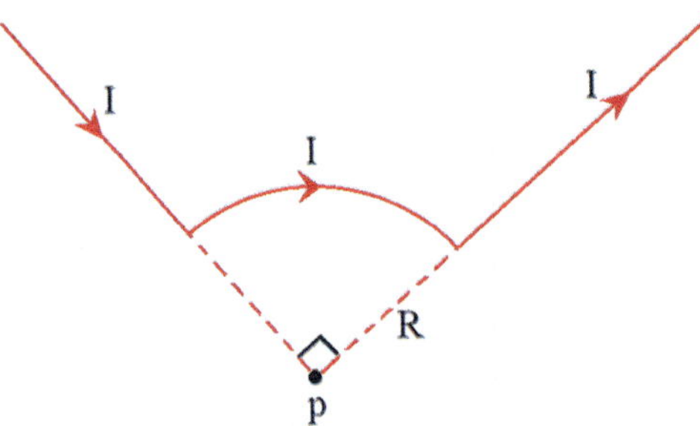

Fig. 5.11 Magnetic field intensity at point p

Solution

As we know, the magnitude of magnetic field intensity (A/m) at the center of a circular arc with angle α, radius r, and current I can be calculated as follows.

$$H = \frac{\alpha}{2\pi} \frac{I}{2r}$$

It should be noted that the two straight pieces of wire do not generate any magnetic field intensity at point p.

Therefore:

$$H = \frac{(\quad)}{(\quad)} \times \frac{(\quad)}{(\quad)}$$

$$\Rightarrow H = \frac{I}{8R}$$

Problem

5.16. In the circuit shown in Fig. 5.12, calculate the magnitude and direction of magnetic field intensity at point p.

Difficulty level ○ Easy ● Normal ○ Hard 

Calculation amount ○ Small ● Normal ○ Large

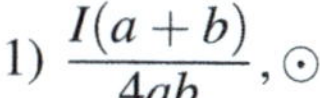

1) $\frac{I(a+b)}{4ab}, \odot$
2) $\frac{I(a+b)}{4ab}, \otimes$
3) $\frac{I(a-b)}{4ab}, \odot$
4) $\frac{I(a-b)}{4ab}, \otimes$

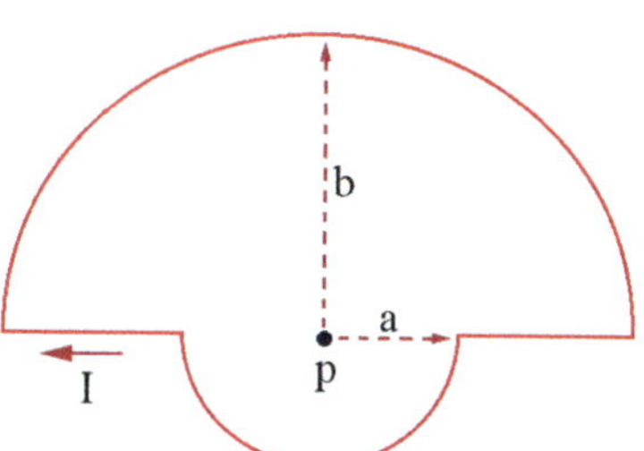

Fig. 5.12 Magnetic field intensity at point p

Partially Solved Exercise

5.17. Calculate the magnitude and direction of magnetic field intensity at point p in the circuit shown in Fig. 5.13.

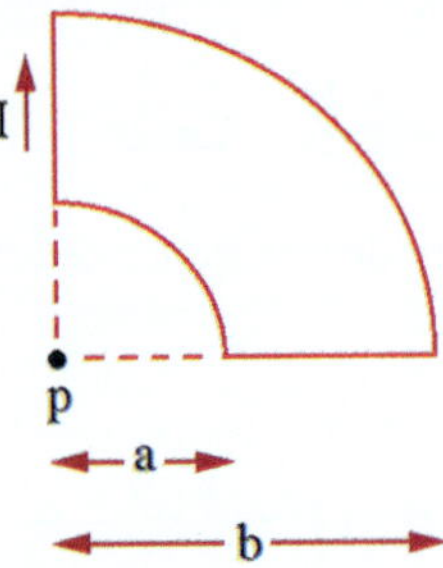

Fig. 5.13 Magnetic field intensity at point p

Final Answer

The two straight pieces of wire do not generate any magnetic field intensity at point p. The direction of magnetic field intensity can be calculated based on the right-hand rule. As is shown in Fig. 5.14, the quarter-circle with radius a generates outward magnetic field intensity while the quarter-circle with radius b generates inward magnetic field intensity.

Moreover, the magnitude of magnetic field intensity (A/m) at the center of a circular arc with angle α, radius r, and current I can be calculated as follows.

$$H = \frac{\alpha}{2\pi}\frac{I}{2r}$$

Hence:

$$H_{tot} = H_1 - H_2$$

$$\Rightarrow H = \frac{(\quad)}{(\quad)} \times \frac{(\quad)}{(\quad)} - \frac{(\quad)}{(\quad)} \times \frac{(\quad)}{(\quad)}$$

$$\Rightarrow H = \frac{I}{8}\left(\frac{1}{a} - \frac{1}{b}\right)$$

$$\Rightarrow H = \frac{I(a-b)}{8ab}$$

The total magnetic field intensity will be outward, that is, ⊙.

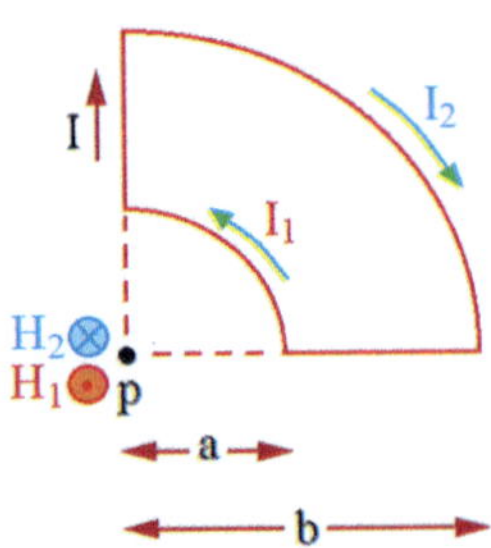

Fig. 5.14 Magnetic field intensity at point p

Problem

5.18. In Fig. 5.15, the current radially enters a circle and leaves it. Calculate the magnitude and direction of magnetic field intensity at the center of the circle.

Difficulty level ○ Easy ○ Normal ● Hard

Calculation amount ○ Small ● Normal ○ Large

1) $\frac{1}{2}\frac{I}{2R}, \odot$
2) $\frac{1}{2}\frac{I}{2R}, \otimes$
3) $\frac{1}{4}\frac{I}{2R}, \otimes$
4) 0

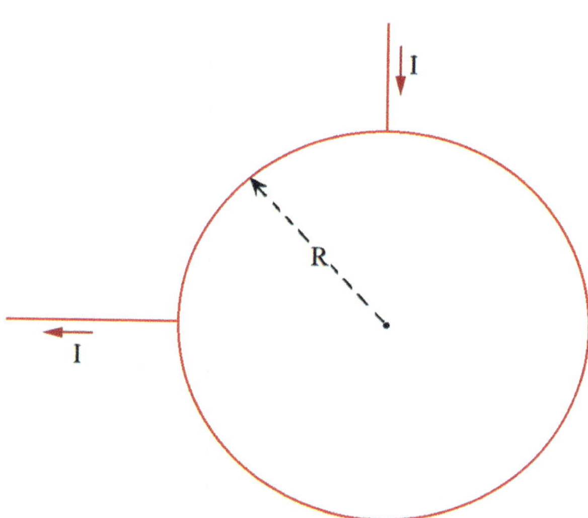

Fig. 5.15 Magnetic field intensity at the center of the circle

5.3 Magnetic Energy and Magnetic Energy Density

Problem

5.19. The amount of a uniform magnetic flux density inside a solenoid is about 1 T. Calculate the magnetic energy density inside the solenoid located in a free space. Herein, assume that $\pi \approx 3$ and $\mu_0 = 4\pi \times 10^{-7}$.

Difficulty level ● Easy ○ Normal ○ Hard

Calculation amount ● Small ○ Normal ○ Large

1) $\omega = 8.33 \times 10^5\ J/m^3$
2) $\omega = 8.33 \times 10^5\ J/m^3$
3) $\omega = 8.33 \times 10^5\ J/m^3$
4) $\omega = 8.33 \times 10^5\ J/m^3$

Problem

5.20. In Problem 5.19, calculate the total magnetic energy stored inside the solenoid if its radius and length are 10 cm and one meter, respectively,

Difficulty level ● Easy ○ Normal ○ Hard

Calculation amount ● Small ○ Normal ○ Large

1) $W = 2.5 \times 10^2\ J$
2) $W = 2.5 \times 10^3\ J$
3) $W = 2.5 \times 10^4\ J$
4) $W = 2.5 \times 10^5\ J$

Exercise

5.21. In an environment, a uniform magnetic flux density with the strength of $\sqrt{2}\ T$ is available. Calculate the magnetic energy density and total magnetic energy stored in a free space with the dimensions of $1 \times 1 \times 1\ m$.

Final Answer

$\omega = \frac{1}{\mu_0}\ J/m^3, W = \frac{1}{\mu_0}\ J$

5.4 Magnetic Force

Problem

5.22. Calculate the magnetic force exerted on an electron that with the velocity of $\vec{v} = 10^6\hat{i} + 1.5 \times 10^6\hat{j}\ m/s$ enters an environment with the magnetic field of $\vec{B} = 0.06\hat{i} - 0.3\hat{j}\ T$. The magnitude of charge of an electron is $1.6 \times 10^{-19}\ C$.

Difficulty level ○ Easy ● Normal ○ Hard

Calculation amount ○ Small ● Normal ○ Large

1) $-6.24 \times 10^{-14}\hat{k}\ N$
2) $6.24 \times 10^{-14}\hat{k}\ N$
3) $6.24 \times 10^{-14}\hat{j}\ N$
4) $-6.24 \times 10^{-14}\hat{j}\ N$

Partially Solved Exercise

5.23. Calculate the magnetic force exerted on a proton that with the velocity of $\vec{v} = 10^6\hat{k}\ m/s$ enters an environment with the magnetic field of $\vec{B} = 2\hat{j}\ T$. The magnitude of charge of a proton is $1.6 \times 10^{-19}\ C$.

Solution

As we know, the magnetic force exerted on a moving charge with the velocity of $\vec{v}$ in an environment with the magnetic field of $\vec{B}$ can be calculated as follows.

$$\vec{F} = q\vec{v} \times \vec{B}$$

Therefore:

$$\vec{F} = (\qquad)(\qquad) \times (\qquad)$$

$$\Rightarrow \vec{F} = (\qquad)(\qquad)$$

$$\Rightarrow \vec{F} = -3.2 \times 10^{-13}\,\hat{i}\,N$$

Notes

In this problem, the relations below have been used.

$$a \times b = (a_1\hat{i} + a_2\hat{j}) \times (b_1\hat{i} + b_2\hat{j}) = (a_1b_2 - a_2b_1)\hat{k}$$

Problem

5.24. Calculate the maximum amount of force exerted on a one-meter wire carrying 2 A current in the uniform magnetic field with $B = 0.1\ T$.

Difficulty level ● Easy ○ Normal ○ Hard

Calculation amount ● Small ○ Normal ○ Large

1) $0.1\ N$
2) $0\ N$
3) $0.4\ N$
4) $0.2\ N$

Problem

5.25. Two long wires carry the currents I_1 and I_2 in the same direction in a free space. The distance between the wires is d. Determine if the force between them is attractive or repulsive. In addition, calculate the magnetic force exerted on one part of each of them with the length of l (Fig. 5.16).

Difficulty level ○ Easy ○ Normal ● Hard

Calculation amount ○ Small ● Normal ○ Large

1) $\frac{\mu_0 l I_1 I_2}{2\pi d}$, repulsive
2) $\frac{\mu_0 l I_1}{2\pi d}$, attractive
3) $\frac{\mu_0 l I_2}{2\pi d}$, repulsive
4) $\frac{\mu_0 l I_1 I_2}{2\pi d}$, attractive

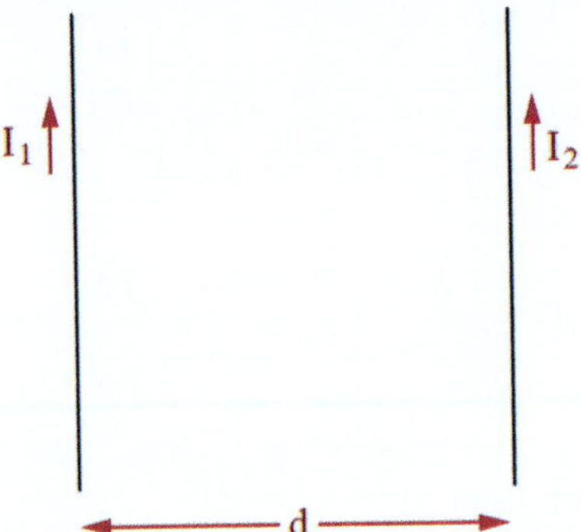

Fig. 5.16 Two parallel current-carrying wires

Problem

5.26. Solve Problem 5.25, while assuming that the currents are in opposite directions (Fig. 5.17).

Difficulty level ○ Easy ○ Normal ● Hard
Calculation amount ○ Small ● Normal ○ Large

1) $\frac{\mu_0 l I_1 I_2}{2\pi d}$, repulsive
2) $\frac{\mu_0 l I_1}{2\pi d}$, attractive
3) $\frac{\mu_0 l I_2}{2\pi d}$, repulsive
4) $\frac{\mu_0 l I_1 I_2}{2\pi d}$, attractive

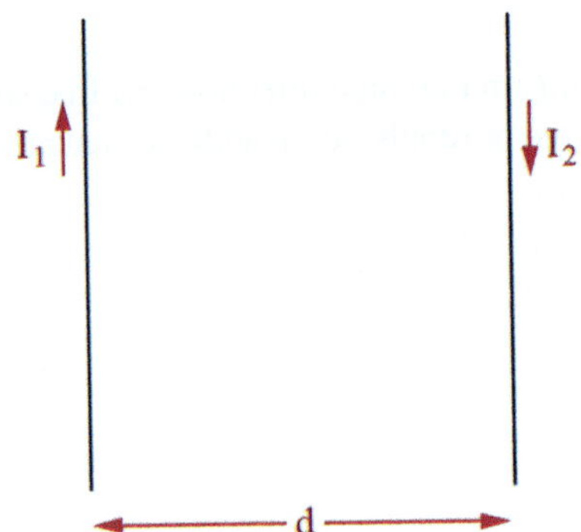

Fig. 5.17 Two parallel current-carrying wires

Exercise

5.27. Two long wires carry the currents $I_1 = 1\ A$ and $I_2 = \pi\ A$ in a free space. The distance between the wires is $0.5\ m$. Calculate the magnetic force that each of the wires exerts on two meters of the other one.

Final Answer

$F = 2\mu_0\ N$

References

1. Rahmani-Andebili, M., General Physics I - Practice Problems, Methods, and Solutions, Springer Nature, 2025.
2. Rahmani-Andebili, M., Calculus III – Practice Problems, Methods, and Solutions, Springer Nature, 2023.
3. Rahmani-Andebili, M., Calculus II – Practice Problems, Methods, and Solutions, Springer Nature, 2023.
4. Rahmani-Andebili, M., Calculus I (2nd Ed.) – Practice Problems, Methods, and Solutions, Springer Nature, 2023.
5. Rahmani-Andebili, M., Precalculus (2nd Ed.) – Practice Problems, Methods, and Solutions, Springer Nature, 2024.

Magnetic Field: Part B

6

Abstract

In this chapter, the problems of the fifth chapter are fully solved, in detail, step-by-step, and with different methods.

6.1 Magnetic Field Intensity Around a Long Current-Carrying Wire

6.1. Based on the information given in the problem, we have [1–5]:

$$r = 1\ m$$

$$B = 4 \times 10^{-7}\ T$$

$$\mu_0 = 4\pi \times 10^{-7}$$

As we know, the magnitude of magnetic field intensity (*A/m*) and magnetic flux density (*T*) around a long wire can be calculated as follows.

$$H = \frac{I}{2\pi r}$$

$$B = \frac{\mu I}{2\pi r}$$

Herein, I and r are the current of the wire and the distance of the point from the wire, respectively. Moreover, $\mu = \mu_0 \mu_r$ is magnetic permeability where μ_r is called relative magnetic permeability which is equal to unity for free space. Additionally, μ_0 is the free space magnetic permeability which is equal to $4\pi \times 10^{-7}$.

Therefore, for this problem, we have:

$$4 \times 10^{-7} = \frac{4\pi \times 10^{-7} I}{2\pi \times 1}$$

$$\Rightarrow I = 2\ A$$

Choice (3) is the answer.

M. Rahmani-Andebili, *General Physics II*, https://doi.org/10.1007/978-3-031-92866-6_6

6.3. Based on the information given in the problem, we have:

$$I_1 = 1\,A$$

$$I_2 = 2\,A$$

$$d = 0.2\,m$$

Figure 6.1 illustrates the magnetic field intensity of each wire at point p based on the right-hand rule. As can be seen, the field intensities are in opposite directions. Therefore:

$$H_{tot} = H_2 - H_1$$

Moreover, the magnitude of magnetic field intensity of each wire can be calculated as follows.

$$H_1 = \frac{I_1}{2\pi r_1} = \frac{1}{2\pi \times 0.1} = \frac{5}{\pi}\,A/m$$

$$H_2 = \frac{I_2}{2\pi r_2} = \frac{2}{2\pi \times 0.1} = \frac{10}{\pi}\,A/m$$

Thus:

$$H_{tot} = \frac{10}{\pi} - \frac{5}{\pi}$$

$$\Rightarrow H_{tot} = \frac{5}{\pi}\,A/m$$

Choice (4) is the answer.

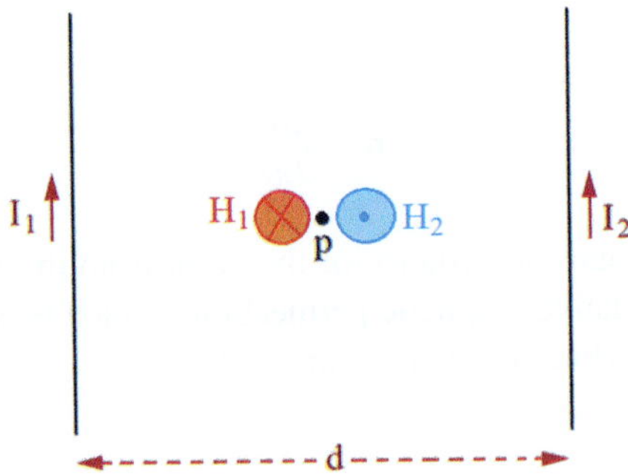

Fig. 6.1 Two parallel current-carrying wires

6.9. Based on the information given in the problem, we have:

$$I_1 = 4\ A$$

$$I_2 = 8\ A$$

$$d_1 = 0.1\ m$$

$$H_p = 0$$

Figure 6.2 illustrates the magnetic field intensity of each wire at point p based on the right-hand rule. As can be seen, the field intensities are in opposite directions. Therefore:

$$H_p = H_2 - H_1$$

Moreover, the magnitude of magnetic field intensity of each wire can be calculated as follows.

$$H_1 = \frac{I_1}{2\pi d_1} = \frac{4}{2\pi \times 0.1} = \frac{20}{\pi}\ A/m$$

$$H_2 = \frac{I_2}{2\pi d_2} = \frac{8}{2\pi d_2} = \frac{4}{\pi d_2}\ A/m$$

Hence:

$$\frac{4}{\pi d_2} - \frac{20}{\pi} = 0$$

$$\Rightarrow \frac{4}{d_2} = 20 \Rightarrow d_2 = 0.2\ m$$

Thus:

$$\Rightarrow d = d_2 + d_1 = 0.2 + 0.1 \Rightarrow d = 0.3\ m$$

Choice (2) is the answer.

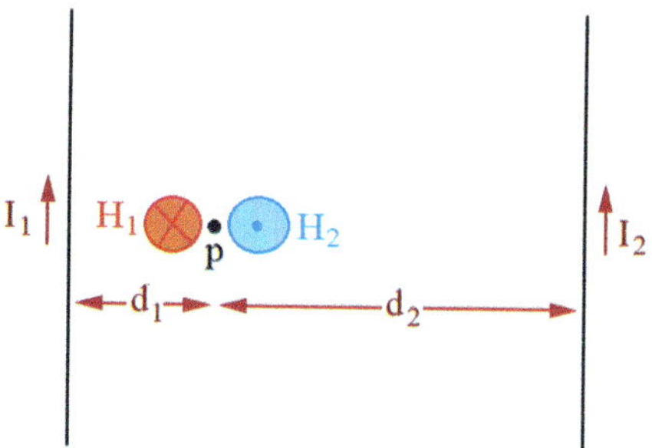

Fig. 6.2 Two parallel current-carrying wires

6.10. Based on the information given in the problem, we have:

$$I_1 = 3\,A$$

$$I_2 = 1\,A$$

$$d_2 = 0.1\,m$$

$$H_p = 0$$

Figure 6.3 illustrates the magnetic field intensity of each wire at point p based on the right-hand rule. As can be seen, the field intensities are in opposite directions. Therefore:

$$H_p = H_2 - H_1$$

Moreover, the magnitude of magnetic field intensity of each wire can be calculated as follows.

$$H_1 = \frac{I_1}{2\pi d_1} = \frac{3}{2\pi \times d_1} = \frac{3}{2\pi d_1}\,A/m$$

$$H_2 = \frac{I_2}{2\pi d_2} = \frac{1}{2\pi \times 0.1} = \frac{5}{\pi}\,A/m$$

Hence:

$$\frac{5}{\pi} - \frac{3}{2\pi d_1} = 0$$

$$\Rightarrow 5 = \frac{3}{2d_1} \Rightarrow d_1 = 0.3\,m$$

Thus:

$$\Rightarrow d = d_1 - d_2 = 0.3 - 0.1 \Rightarrow d = 0.2\,m$$

Choice (1) is the answer.

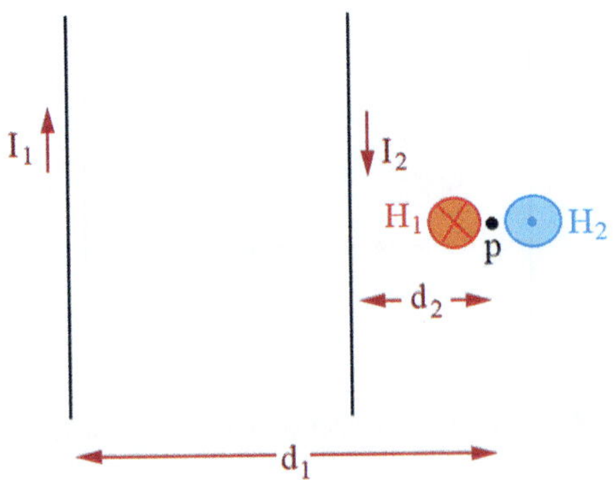

Fig. 6.3 Two parallel current-carrying wires

6.11. Based on the information given in the problem, we have:

$$I_1 = 3\,A$$

$$I_2 = 4\,A$$

$$d = 2\,m$$

Figure 6.4 illustrates the magnetic field intensity of each wire at point p based on the right-hand rule. The distance of the point from each wire can be calculated as follows.

$$\cos 45^\circ = \frac{a}{d}$$

$$\Rightarrow \frac{\sqrt{2}}{2} = \frac{a}{d} \Rightarrow a = \frac{d}{\sqrt{2}}$$

The magnitude of magnetic field intensity of each wire at point p can be calculated as follows.

$$H_1 = \frac{I_1}{2\pi a} = \frac{I_1}{2\pi\left(\frac{d}{\sqrt{2}}\right)} = \frac{\sqrt{2}}{2\pi d} I_1$$

$$H_2 = \frac{I_2}{2\pi a} = \frac{I_2}{2\pi\left(\frac{d}{\sqrt{2}}\right)} = \frac{\sqrt{2}}{2\pi d} I_2$$

The angle between the vectors of magnetic field intensity is 90°. The magnitude of resultant magnetic field intensity can be calculated by using the Pythagorean formula as follows.

$$H = \sqrt{H_1^2 + H_2^2}$$

$$\Rightarrow H = \frac{\sqrt{2}}{2\pi d}\sqrt{I_1^2 + I_2^2}$$

$$\Rightarrow H = \frac{\sqrt{2}}{2\pi \times 2}\sqrt{3^2 + 4^2}$$

$$\Rightarrow H = \frac{5}{2\sqrt{2}\pi}\,A/m$$

Choice (1) is the answer.

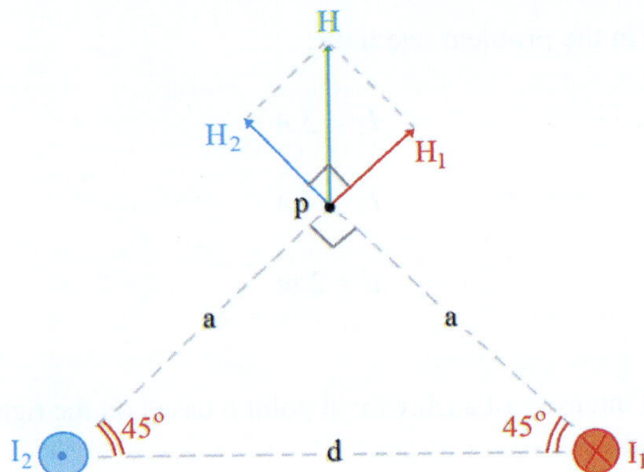

Fig. 6.4 The magnetic field intensity of two parallel current-carrying wires at point p

6.2 Magnetic Field Intensity at the Center of a Circle and a Circular Arc

6.12. Based on the information given in the problem, we have:

$$\mu_r = 1$$

$$r = 0.6\ m$$

$$I = 10\ A$$

$$\pi \approx 3$$

$$\mu_0 = 4\pi \times 10^{-7}$$

As we know, the magnitude of magnetic field intensity (A/m) and magnetic flux density (T) at the center of a circle with radius r carrying current I can be calculated as follows.

$$H = \frac{I}{2r}$$

$$B = \frac{\mu I}{2r}$$

Herein, $\mu = \mu_0 \mu_r$ is magnetic permeability where μ_r is called relative magnetic permeability which is equal to unity for free space. Additionally, μ_0 is the free space magnetic permeability which is equal to $4\pi \times 10^{-7}$.
Thus:

$$B = \frac{\mu_0 I}{2r} = \frac{4\pi \times 10^{-7} \times 10}{2 \times 0.6} = \frac{4 \times 3 \times 10^{-7} \times 10}{2 \times 0.6}$$

$$\Rightarrow B = 10^{-5}\ T$$

Choice (4) is the answer.

6.14. The magnitude of magnetic field intensity (A/m) at the center of a circular arc with angle α, radius r, and current I can be calculated as follows.

$$H = \frac{\alpha}{2\pi}\frac{I}{2r}$$

Hence:

$$H = \frac{\theta I}{4\pi R}$$

Choice (4) is the answer (Fig. 6.5).

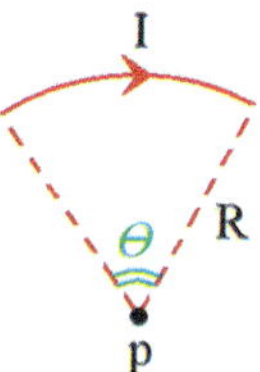

Fig. 6.5 Magnetic field intensity at the center of a circular arc

---

6.16. The two straight pieces of wire do not generate any magnetic field intensity at point p. The direction of magnetic field intensity can be calculated based on the right-hand rule. Both half-circles generate inward magnetic field intensity shown in Fig. 6.6. Thus, the total magnetic field intensity will be inward, that is, $\otimes$.

Moreover, the magnitude of magnetic field intensity (A/m) at the center of a circular arc with angle α, radius r, and current I can be calculated as follows.

$$H = \frac{\alpha}{2\pi}\frac{I}{2r}$$

Hence:

$$H_{tot} = H_1 + H_2$$

$$H = \frac{\pi}{2\pi}\frac{I}{2a} + \frac{\pi}{2\pi}\frac{I}{2b} \Rightarrow H = \frac{I}{4}\left(\frac{1}{a} + \frac{1}{b}\right)$$

$$\Rightarrow H = \frac{I(a+b)}{4ab}$$

Choice (2) is the answer.

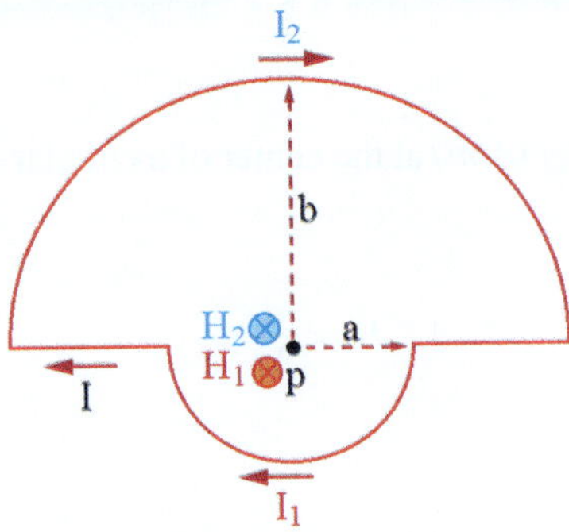

Fig. 6.6 Magnetic field intensity at point p

6.18. First, the current of each part of the circle needs to be calculated as follows. Based on current division rule, we can write:

$$I_1 = \frac{R_2}{R} I$$

$$I_2 = \frac{R_1}{R} I$$

Where, R is the total resistance of the circle, R_1 is the resistance of the quarter-circle, and R_2 is the resistance of the rest of the circle. Therefore:

$$I_1 = \frac{3}{4} I$$

$$I_2 = \frac{1}{4} I$$

The two straight pieces of wire do not generate any magnetic field intensity at point p. The direction of magnetic field intensity can be calculated based on the right-hand rule. As is shown in Fig. 6.7, the quarter-circle generates outward magnetic field intensity while the other part of the circle generates inward magnetic field intensity.

Moreover, the magnitude of magnetic field intensity (A/m) at the center of a circular arc with angle α, radius r, and current I can be calculated as follows.

$$H = \frac{\alpha}{2\pi} \frac{I}{2r}$$

Hence:

$$H_{tot} = H_1 - H_2$$

$$H = \frac{\frac{\pi}{2}}{2\pi}\frac{I_1}{2R} - \frac{\frac{3\pi}{2}}{2\pi}\frac{I_2}{2R}$$

$$\Rightarrow H = \frac{I_1}{8R} - \frac{3I_2}{8R}$$

$$\Rightarrow H = \frac{\frac{3}{4}I}{8R} - \frac{3 \times \frac{1}{4}I}{8R}$$

$$\Rightarrow H = \frac{3I}{32R} - \frac{3I}{32R}$$

$$\Rightarrow H = 0$$

Choice (4) is the answer.

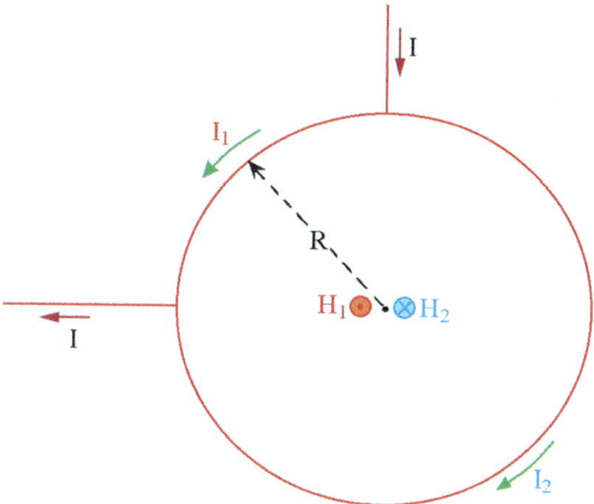

Fig. 6.7 Magnetic field intensity at the center of the circle

6.3 Magnetic Energy and Magnetic Energy Density

6.19. Based on the information given in the problem, we have:

$$B = 1\,T$$

$$\mu_r = 1$$

$$\pi \approx 3$$

$$\mu_0 = 4\pi \times 10^{-7}$$

As we know, the magnetic energy density (J/m^3) in a space can be calculated as follows.

$$\omega = \frac{1}{2}\mu H^2 = \frac{1}{2}\mu BH = \frac{1}{2}\frac{B^2}{\mu}$$

Therefore:

$$\omega = \frac{1}{2}\frac{B^2}{\mu_r \mu_0}$$

$$\Rightarrow \omega = \frac{1}{2}\frac{1^2}{1 \times 4\pi \times 10^{-7}} = \frac{1}{4 \times 3 \times 10^{-7}}$$

$$\Rightarrow \omega = 8.33 \times 10^5 \ J/m^3$$

Choice (1) is the answer.

6.20. Based on the information given in the problem, we have:

$$r = 0.1 \ m$$

$$l = 1 \ m$$

$$\pi \approx 3$$

$$\omega = 8.33 \times 10^5 \ J/m^3$$

The total magnetic energy (J) in a space can be calculated as follows.

$$W = \omega V$$

Hence:

$$W = \omega l A = \omega l \pi r^2$$

$$\Rightarrow W = 8.33 \times 10^5 \times 1 \times \pi \times 0.1^2$$

$$\Rightarrow W = 2.5 \times 10^4 \ J$$

Choice (3) is the answer.

6.4 Magnetic Force

6.22. Based on the information given in the problem, we have:

$$\vec{v} = 10^6\,\hat{i} + 1.5 \times 10^6\,\hat{j}\;m/s$$

$$\vec{B} = 0.06\,\hat{i} - 0.3\,\hat{j}\;T$$

$$q = -1.6 \times 10^{-19}\;C$$

As we know, the magnetic force exerted on a moving charge with the velocity of $\vec{v}$ in an environment with the magnetic field of $\vec{B}$ can be calculated as follows.

$$\vec{F} = q\,\vec{v} \times \vec{B}$$

Therefore:

$$\vec{F} = (-1.6 \times 10^{-19})\left(10^6\,\hat{i} + 1.5 \times 10^6\,\hat{j}\right) \times \left(0.06\,\hat{i} - 0.3\,\hat{j}\right)$$

$$\Rightarrow \vec{F} = (-1.6 \times 10^{-19})\left(-39 \times 10^4\,\hat{k}\right)$$

$$\Rightarrow \vec{F} = 6.24 \times 10^{-14}\,\hat{k}\;N$$

Choice (2) is the answer.

Notes

In this problem, the relation below has been used.

$$a \times b = \left(a_1\,\hat{i} + a_2\,\hat{j}\right) \times \left(b_1\,\hat{i} + b_2\,\hat{j}\right) = (a_1 b_2 - a_2 b_1)\,\hat{k}$$

6.24. Based on the information given in the problem, we have:

$$l = 1\;m$$

$$I = 2\;A$$

$$B = 0.1\;T$$

The magnetic force exerted on a current-carrying wire with the current and length of I and l in an environment with the magnetic field of $\vec{B}$ can be calculated as follows.

$$\vec{F} = I\vec{l} \times \vec{B} = BIl \sin\theta$$

where, θ is the angle between the $\vec{l}$ and $\vec{B}$.

As can be noticed from the relation, the maximum amount of force can be achieved for $\theta = \frac{\pi}{2}$. Therefore:

$$F_{max} = BIl$$

$$\Rightarrow F_{max} = (0.1)(2)(1)$$

$$\Rightarrow F_{max} = 0.2\ N$$

Choice (4) is the answer.

6.25. As we know, the magnetic force exerted on a current-carrying wire with the current and length of I and l in an environment with the magnetic field of $\vec{B}$ can be calculated as follows.

$$\vec{F} = I\vec{l} \times \vec{B} = BIl \sin\theta$$

where, θ is the angle between the $\vec{l}$ and $\vec{B}$.

The magnitude of magnetic flux density of the first wire on the place of the second wire can be calculated as follows. As can be seen in Fig. 6.8 (a), it is inward.

$$B_1 = \frac{\mu_0 I_1}{2\pi d}$$

Thus, the magnitude of magnetic force that the first wire exerts on the second one can be calculated as follows. Herein, $\theta = 90°$, as can be noticed from Fig. 6.8 (a).

$$F_{12} = \left(\frac{\mu_0 I_1}{2\pi d}\right)(I_2)(l) \sin 90°$$

$$\Rightarrow F_{12} = \frac{\mu_0 I_1 I_2 l}{2\pi d}$$

The direction of magnetic force that the first wire exerts on the second one is determined based on the right-hand rule shown in Fig. 6.8 (a). As can be seen, it is toward the first wire.

On the other hand, the magnitude of magnetic flux density of the second wire on the place of the first one can be calculated as follows. As can be seen in Fig. 6.8 (b), it is outward.

$$B_2 = \frac{\mu_0 I_2}{2\pi d}$$

Hence, the magnitude of magnetic force that the second wire exerts on the first one can be calculated as follows. Herein, $\theta = 90°$, as can be noticed from Fig. 6.8 (b).

$$F_{21} = \left(\frac{\mu_0 I_2}{2\pi d}\right)(I_1)(l)\sin 90°$$

$$\Rightarrow F_{21} = \frac{\mu_0 I_1 I_2 l}{2\pi d}$$

The direction of magnetic force that the second wire exerts on the first one is determined based on the right-hand rule shown in Fig. 6.8 (b). As can be seen, it is toward the second wire.

As can be noticed from Fig. 6.8 (a) and (b), the wires attract each other. Moreover, we have:

$$F_{12} = F_{21} = \frac{\mu_0 I_1 I_2 l}{2\pi d}$$

Choice (4) is the answer.

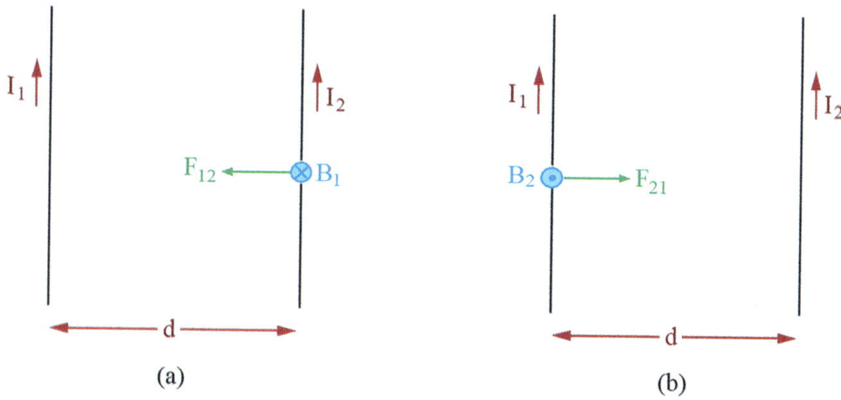

Fig. 6.8 (**a**) The force that the first wire exerts on the second one. (**b**) The force that the second wire exerts on the first one

6.26. As we know, the magnetic force exerted on a current-carrying wire with the current and length of I and l in an environment with the magnetic field of $\vec{B}$ can be calculated as follows.

$$\vec{F} = I\,\vec{l} \times \vec{B} = BIl\sin\theta$$

where, θ is the angle between the $\vec{l}$ and $\vec{B}$.

The magnitude of magnetic flux density of the first wire on the place of the second wire can be calculated as follows. As can be seen in Fig. 6.9 (a), it is inward.

$$B_1 = \frac{\mu_0 I_1}{2\pi d}$$

Thus, the magnitude of magnetic force that the first wire exerts on the second one can be calculated as follows. Herein, $\theta = 90°$, as can be noticed from Fig. 6.9 (a).

$$F_{12} = \left(\frac{\mu_0 I_1}{2\pi d}\right)(I_2)(l)\sin 90°$$

$$\Rightarrow F_{12} = \frac{\mu_0 I_1 I_2 l}{2\pi d}$$

The direction of magnetic force that the first wire exerts on the second one is determined based on the right-hand rule shown in Fig. 6.9 (a). As can be seen, it is away from the first wire.

On the other hand, the magnitude of magnetic flux density of the second wire on the place of the first one can be calculated as follows. As can be seen in Fig. 6.9 (b), it is inward.

$$B_2 = \frac{\mu_0 I_2}{2\pi d}$$

Hence, the magnitude of magnetic force that the second wire exerts on the first one can be calculated as follows. Herein, $\theta = 90°$, as can be noticed from Fig. 6.9 (b).

$$F_{21} = \left(\frac{\mu_0 I_2}{2\pi d}\right)(I_1)(l)\sin 90°$$

$$\Rightarrow F_{21} = \frac{\mu_0 I_1 I_2 l}{2\pi d}$$

The direction of magnetic force that the second wire exerts on the first one is determined based on the right-hand rule shown in Fig. 6.9 (b). As can be seen, it is away from the second wire.

As can be noticed from Fig. 6.9 (a) and (b), the wires repel each other. Moreover, we have:

$$F_{12} = F_{21} = \frac{\mu_0 I_1 I_2 l}{2\pi d}$$

Choice (1) is the answer.

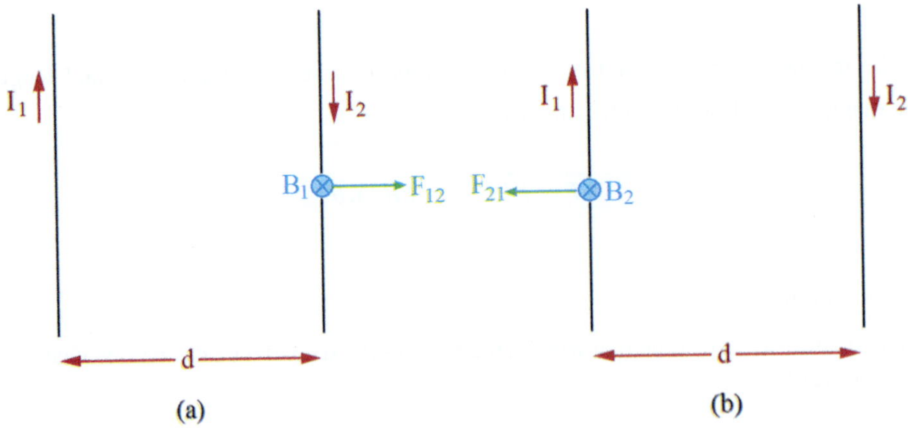

Fig. 6.9 (**a**) The force that the first wire exerts on the second one. (**b**) The force that the second wire exerts on the first one

References

1. Rahmani-Andebili, M., General Physics I - Practice Problems, Methods, and Solutions, Springer Nature, 2025.
2. Rahmani-Andebili, M., Calculus III – Practice Problems, Methods, and Solutions, Springer Nature, 2023.
3. Rahmani-Andebili, M., Calculus II – Practice Problems, Methods, and Solutions, Springer Nature, 2023.
4. Rahmani-Andebili, M., Calculus I (2nd Ed.) – Practice Problems, Methods, and Solutions, Springer Nature, 2023.
5. Rahmani-Andebili, M., Precalculus (2nd Ed.) – Practice Problems, Methods, and Solutions, Springer Nature, 2024.

Electromagnetic Induction: Part A

7

Abstract

In this chapter, the basic and advanced problems of Electromagnetic Induction are studied. The subjects include Magnetic Flux, Voltage Induced in a Coil, Voltage Induced in a Wire, Inductance and Equivalent Inductance of a Circuit, Magnetic Energy Density, and Magnetic Energy Stored in Inductors. Herein, different types of problems and exercises are presented that are categorized as follows.

- ***Problems with detailed solution***: They have been designed to teach students the subjects in detail. Moreover, they have been categorized in different levels based on their difficulty levels (easy, normal, and hard) and calculation amounts (small, normal, and large).
- ***Partially solved exercises***: They have been designed to encourage students to practice problems while guiding them through the problem-solving procedure and hinting the required formulas.
- ***Exercises with final answer***: They have been designed to encourage students to practice more by themselves while hinting them by the final answer as well as to help instructors to give tests or quizzes.

7.1 Magnetic Flux

Problem

7.1. Figure 7.1 illustrates a current-carrying wire which is perpendicular to the plain of the ring. Calculate the amount of magnetic flux (φ) passing through the ring [1–5].

Difficulty level ● Easy ○ Normal ○ Hard 

Calculation amount ● Small ○ Normal ○ Large

1) $\frac{\mu_0 I a}{2}$
2) $\frac{\mu_0 I}{2\pi a}$
3) $\frac{Ia}{2}$
4) 0

M. Rahmani-Andebili, *General Physics II*, https://doi.org/10.1007/978-3-031-92866-6_7

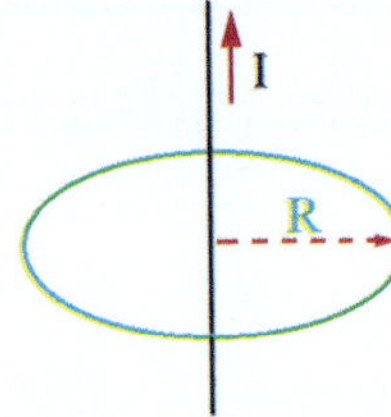

Fig. 7.1 A wire perpendicularly passing through the center of the circle

7.2 Voltage Induced in a Coil

Problem

7.2. Calculate the voltage induced in a coil including ten turns placed in a magnetic field with the magnetic flux equation of $\varphi(t) = t\ wb$.

Difficulty level ● Easy ○ Normal ○ Hard

Calculation amount ● Small ○ Normal ○ Large

1) 10 V
2) 0 V
3) 1 V
4) $-10\ V$

Partially Solved Exercise

7.3. Calculate the maximum voltage across each turn of a coil placed in a magnetic field with the magnetic flux equation of $\varphi(t) = \sin 10t\ wb$.

Solution

Based on the information given in the problem, we have:

$$N = 1$$

$$\varphi(t) = \sin 10t\ wb$$

As we know, the instantaneous voltage induced ($e(t)$) across a coil, inductor, or solenoid with N turns placed in a magnetic field with the magnetic flux of $\varphi(t)$ can be calculated as follows.

$$e(t) = -N\frac{d\varphi(t)}{dt}$$

Moreover, the average voltage induced ($\overline{e}$) across the coil can be calculated as follows.

$$\overline{e} = -N\frac{\Delta\varphi}{\Delta t} = -N\frac{\varphi_2 - \varphi_1}{t_2 - t_1}$$

Thus:

$$e(t) = -(\quad) \times \frac{d}{dt}(\quad)$$

$$\Rightarrow e(t) = -(\quad)\cos 10t$$

$$\Rightarrow |e_{max}| = 10\ V$$

Notes

In this problem, the relations below have been used.

$$\frac{d}{dt}(\sin at) = a\cos at$$

$$-1 \leq \cos at \leq 1$$

Problem

7.4. The primary radius of a ring is about one meter, but it starts decreasing with the rate of 10 cm/s. If a uniform magnetic field with the strength of 100 T is perpendicularly applied to the ring, calculate the induced voltage in it.

Difficulty level ○ Easy ● Normal ○ Hard

Calculation amount ○ Small ● Normal ○ Large

1) $-20\ V$
2) $-200\pi\ V$
3) $200\ V$
4) $-20\pi\ V$

Partially Solved Exercise

7.5. The magnetic flux of $\varphi(t) = 5\sin\left(100t - \frac{\pi}{4}\right)$ is passing from a 10-turn and 10-Ω coil. Calculate the maximum value of the current induced in the ring.

Solution

Based on the information given in the problem, we have:

$$\varphi(t) = 5\sin\left(100t - \frac{\pi}{4}\right)$$

$$N = 10$$

$$R = 10\ \Omega$$

As we know, the instantaneous voltage induced ($e(t)$) in a coil, inductor, or solenoid with N turns placed in a magnetic field with the magnetic flux of $\varphi(t)$ can be calculated as follows.

$$e(t) = -N\frac{d\varphi(t)}{dt}$$

Moreover, the average voltage induced ($\overline{e}$) across the coil can be calculated as follows.

$$\bar{e} = -N\frac{\Delta\varphi}{\Delta t} = -N\frac{\varphi_2 - \varphi_1}{t_2 - t_1}$$

Hence:

$$e(t) = -(\qquad) \times \frac{d}{dt}(\qquad\qquad\qquad)$$

$$\Rightarrow e(t) = -(\qquad)\cos\left(100t - \frac{\pi}{4}\right)$$

$$\Rightarrow |e_{max}| = (\quad)\ V$$

Using Ohm's law, we have:

$$|I_{max}| = \frac{|e_{max}|}{R} = \frac{(\quad)}{(\quad)}$$

$$\Rightarrow |I_{max}| = 500\ A$$

Notes

In this problem, the relations below have been used.

$$\frac{d}{dt}(\sin(at + \theta)) = a\cos(at + \theta)$$

$$-1 \leq \cos(at + \theta) \leq 1$$

Problem

7.6. Calculate the amount of voltage induced in each turn of a coil that its magnetic flux changes from $-0.1\ wb$ to $0.9\ wb$ during $0.125\ s$.

Difficulty level ● Easy ○ Normal ○ Hard

Calculation amount ● Small ○ Normal ○ Large

1) $-8\ V$
2) $8\ V$
3) $1\ V$
4) $-1\ V$

Problem

7.7. Figure 7.2 shows the magnetic flux-time curve in one part of the space. Determine the induced voltage-time curve of a ring perpendicularly placed in this environment.

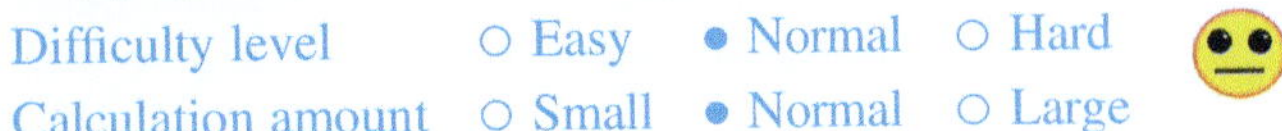

Difficulty level ○ Easy ● Normal ○ Hard

Calculation amount ○ Small ● Normal ○ Large

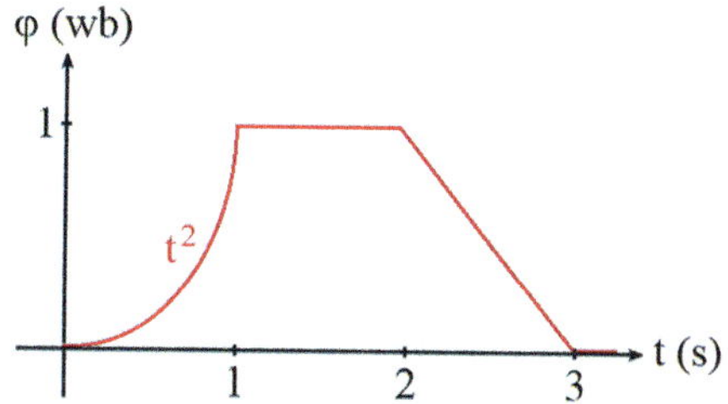

Fig. 7.2 The magnetic flux-time curve in a space

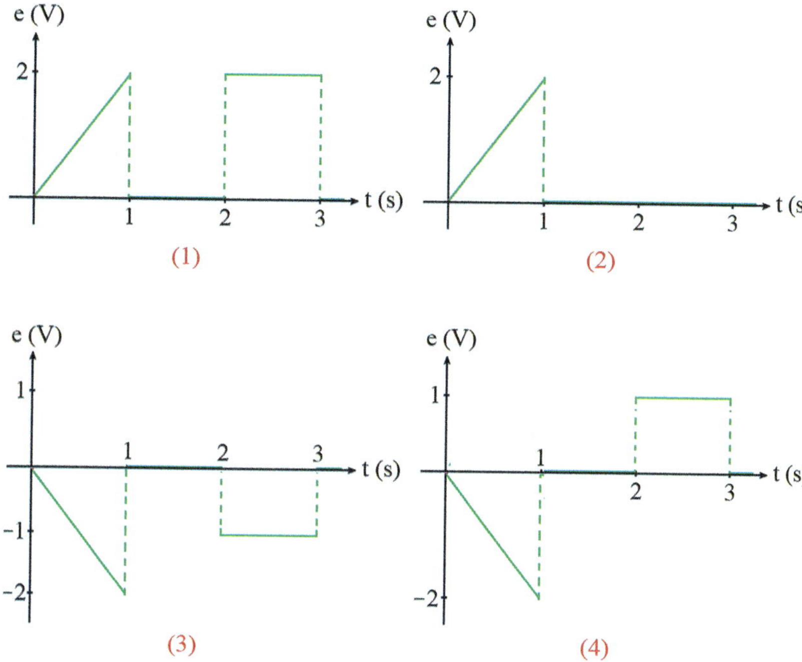

7.3 Voltage Induced in a Wire

Problem

7.8. A 2-m straight wire moves at the velocity of 20 m/s in a 1-T magnetic field. If the movement direction of the wire is perpendicular to the field, calculate the voltage induced in that.

Difficulty level ● Easy ○ Normal ○ Hard

Calculation amount ● Small ○ Normal ○ Large

1) 10 V
2) 20 V
3) 40 V
4) 2 V

Partially Solved Exercise

7.9. Figure 7.3 shows a circuit in which the rod AB with the length of 0.5 m is moving to the right at $v = 5\ m/s$ in presence of the inward magnetic field of $B = 2\ T$. The resistance of the circuit is zero; however, the rod has 3 Ω resistance. Calculate the current included in the circuit.

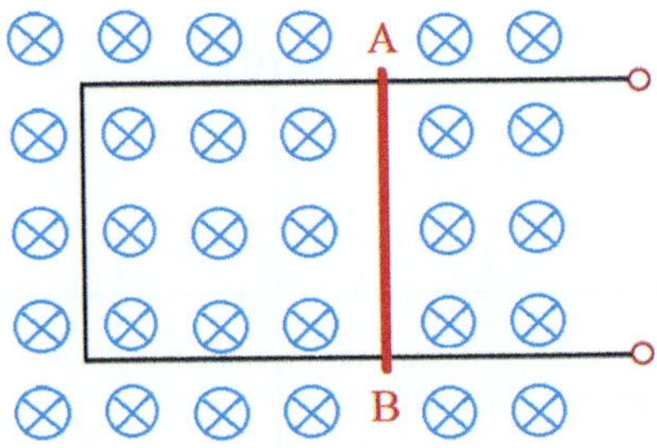

Fig. 7.3 A rod moving to the right in the presence of a magnetic field

Solution

Based on the information given in the problem, we have:

$$l = 0.5\ m$$

$$v = 5\ m/s$$

$$B = 2\ T$$

$$\theta = 90^\circ$$

$$R = 3\ \Omega$$

As we know, the voltage induced in a straight wire with the length of l that moves at velocity $\vec{v}$ in an environment with the magnetic field of $\vec{B}$ can be calculated as follows.

$$e = l\vec{v} \times \vec{B} = Blv\sin\theta$$

Herein, θ is the angle between the $\vec{v}$ and $\vec{B}$.

Hence:

$$e = (\quad) \times (\quad) \times (\quad) \sin(\quad)^\circ$$

$$\Rightarrow e = (\quad)\ V$$

Using Ohm's law, we have:

$$I = \frac{e}{R}$$

$$\Rightarrow I = \frac{5}{3}\ A$$

7.4 Inductance and Equivalent Inductance of a Circuit

Problem

7.10. In an experiment, the current of a solenoid decreases from 10 A to zero for two seconds while a 5 V voltage is induced in it. Calculate the inductance of the solenoid.

Difficulty level ● Easy ○ Normal ○ Hard

Calculation amount ● Small ○ Normal ○ Large

1) 0.5 H
2) 1 H
3) 5 H
4) 0.2 H

Exercise

7.11. The current of a 0.01 $- H$ solenoid is $i(t) = \sin\left(100t - \frac{\pi}{4}\right)$ A. Calculate the voltage induced in it.

Final Answer

$e(t) = -\cos\left(100t - \frac{\pi}{4}\right)$

Problem

7.12. In the circuit shown in Fig. 7.4, the inductors are connected in series. Calculate the equivalent inductance of the circuit.

Difficulty level ● Easy ○ Normal ○ Hard

Calculation amount ● Small ○ Normal ○ Large

1) 11 H
2) 1 H
3) 4 H
4) 2 H

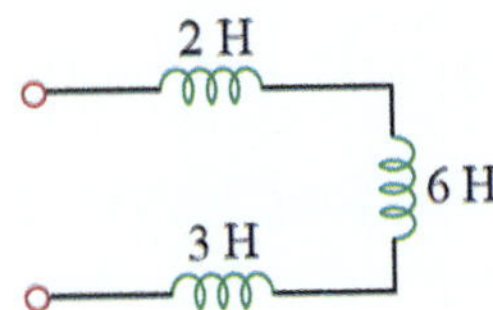

Fig. 7.4 The series connection of inductors

Problem

7.13. In the circuit shown in Fig. 7.5, the inductors are connected in parallel. Calculate the equivalent inductance of the circuit.

Difficulty level	● Easy	○ Normal	○ Hard
Calculation amount	● Small	○ Normal	○ Large

1) 11 H
2) 1 H
3) 4 H
4) 2 H

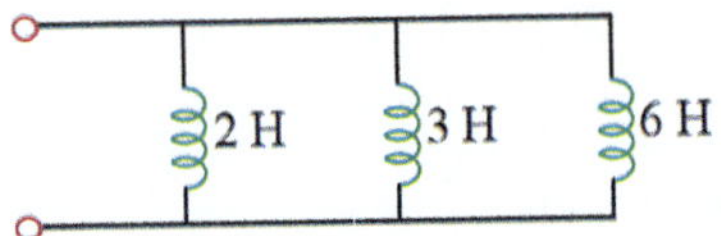

Fig. 7.5 The parallel connection of inductors

Problem

7.14. In the circuit shown in Fig. 7.6, the inductors are connected in series-parallel from. Calculate the equivalent inductance of the circuit.

Difficulty level	● Easy	○ Normal	○ Hard
Calculation amount	● Small	○ Normal	○ Large

1) 1 H
2) 2 H
3) 4 H
4) 2.5 H

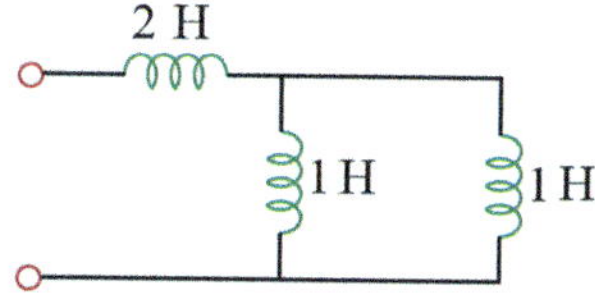

Fig. 7.6 The series-parallel connection of inductors

Partially Solved Exercise

7.15. Calculate the equivalent inductance of the circuit shown in Fig. 7.7.

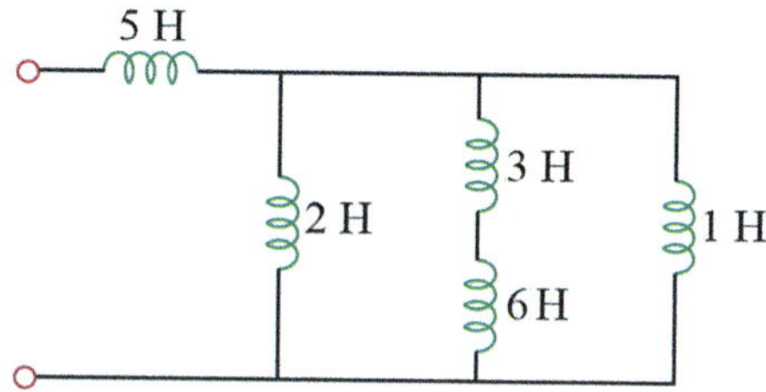

Fig. 7.7 The series-parallel connection of inductors

Solution

The equivalent inductance of the circuit can be calculated as follows.

First, the series connection of the middle branch should be simplified as follows.

$$L_{3,6} = L_3 + L_6 = (\quad) + (\quad) = 9\,H$$

Then, the three parallel branches can be combined as follows.

$$\frac{1}{L_{2,9,1}} = \frac{1}{L_2} + \frac{1}{L_9} + \frac{1}{L_1}$$

$$\Rightarrow \frac{1}{L_{2,9,1}} = \frac{1}{(\quad)} + \frac{1}{(\quad)} + \frac{1}{(\quad)} = \frac{(\quad) + (\quad) + (\quad)}{18} = \frac{29}{18}$$

$$\Rightarrow L_{2,9,1} = \frac{18}{29}\,H$$

Finally, there are only two inductors connected in series. Thus:

$$L_{eq} = (\quad) + (\quad) = \frac{(\quad) + (\quad)}{(\quad)}$$

$$\Rightarrow L_{eq} = \frac{163}{29}\,H$$

7.5 Magnetic Energy Stored in Inductors

Problem

7.16. The current of a $2-H$ inductor is $3\ A$. Calculate the magnetic energy stored in it.

Difficulty level ○ Easy ○ Normal ○ Hard

Calculation amount ○ Small ○ Normal ○ Large

1) $18\ J$
2) $9\ J$
3) $3\ J$
4) $6\ J$

Partially Solved Exercise

7.17. The current of a series circuit, including three inductors, that is, $1\ H$, $2\ H$, and $3\ H$ is one Ampere. Calculate the total magnetic energy stored in them.

Solution

The magnetic energy stored in an inductor can be calculated as follows.

$$W=\frac{1}{2}LI^2$$

Moreover, the equivalent inductance of the circuit can be calculated as follows.

$$L_{eq}=1+2+3=(\quad)\ H$$

Therefore:

$$W=\frac{1}{2}\times(\quad)\times(\quad)^2$$

$$\Rightarrow W=3\ J$$

Partially Solved Exercise

7.18. The current of a parallel circuit, including two inductors, that is, $1\ H$ and $2\ H$ is one Ampere. Calculate the total magnetic energy stored in them.

Solution

The magnetic energy stored in an inductor can be calculated as follows.

$$W=\frac{1}{2}LI^2$$

Moreover, the equivalent inductance of the circuit can be calculated as follows.

$$L_{eq}=\frac{1\times 2}{1+2}=(\quad)\ H$$

Therefore:

$$W = \frac{1}{2} \times (\quad) \times (\quad)^2$$

$$\Rightarrow W = \frac{1}{3}\ J$$

References

1. Rahmani-Andebili, M., General Physics I - Practice Problems, Methods, and Solutions, Springer Nature, 2025.
2. Rahmani-Andebili, M., Calculus III – Practice Problems, Methods, and Solutions, Springer Nature, 2023.
3. Rahmani-Andebili, M., Calculus II – Practice Problems, Methods, and Solutions, Springer Nature, 2023.
4. Rahmani-Andebili, M., Calculus I (2nd Ed.) – Practice Problems, Methods, and Solutions, Springer Nature, 2023.
5. Rahmani-Andebili, M., Precalculus (2nd Ed.) – Practice Problems, Methods, and Solutions, Springer Nature, 2024.

Electromagnetic Induction: Part B

8

Abstract
In this chapter, the problems of the seventh chapter are fully solved, in detail, step-by-step, and with different methods.

8.1 Magnetic Flux

8.1. Based on the right-hand rule, the magnetic field of the wire are closed-pass concentric circles centered around the wire shown in Fig. 8.1 [1–5]. This magnetic field is tangent to the surface of the circle. Therefore, no magnetic flux passes through the circle.

Choice (4) is the answer.

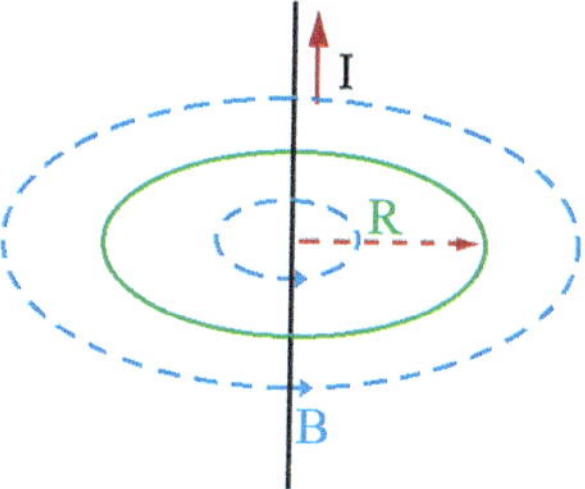

Fig. 8.1 A wire perpendicularly passing through the center of the circle

8.2 Voltage Induced in a Coil

8.2. Based on the information given in the problem, we have:

$$N = 10$$

$$\varphi(t) = t \; wb$$

M. Rahmani-Andebili, *General Physics II*, https://doi.org/10.1007/978-3-031-92866-6_8

The instantaneous voltage induced ($e(t)$) in a coil, inductor, or solenoid with N turns placed in a magnetic field with the magnetic flux of $\varphi(t)$ can be calculated as follows.

$$e(t) = -N\frac{d\varphi(t)}{dt}$$

Moreover, the average voltage induced ($\overline{e}$) across the coil can be calculated as follows.

$$\overline{e} = -N\frac{\Delta\varphi}{\Delta t} = -N\frac{\varphi_2 - \varphi_1}{t_2 - t_1}$$

Therefore:

$$e(t) = -10\frac{d}{dt}(t)$$

$$e(t) = -10\ V$$

Choice (4) is the answer.

Notes

In this problem, the relation below has been used.

$$\frac{d}{dt}t^n = nt^{n-1}$$

8.4. Based on the information given in the problem, we have:

$$N = 1$$

$$r = 1\ m$$

$$\frac{dr}{dt} = 0.1\ m/s$$

$$B = 100\ T$$

As we know, the instantaneous voltage induced ($e(t)$) in a coil, inductor, or solenoid with N turns placed in a magnetic field with the magnetic flux of $\varphi(t)$ can be calculated as follows.

$$e(t) = -N\frac{d\varphi(t)}{dt}$$

Moreover, the average voltage induced ($\overline{e}$) in the coil can be calculated as follows.

$$\overline{e} = -N\frac{\Delta\varphi}{\Delta t} = -N\frac{\varphi_2 - \varphi_1}{t_2 - t_1}$$

Hence:

$$e(t) = -N \times \frac{d}{dt}(BA)$$

$$\Rightarrow e(t) = -N \times \frac{d}{dt}(B\pi r^2)$$

$$\Rightarrow e(t) = -NB\pi\left(2r\frac{dr}{dt}\right)$$

$$\Rightarrow e(t) = -1 \times 100\pi(2 \times 1 \times 0.1)$$

$$\Rightarrow e(t) = -20\pi \ V$$

Choice (4) is the answer.

Notes

In this problem, the relations below have been used.

$$\varphi = BA$$

$$\frac{d}{dt}r^n = nr^{n-1}\frac{dr}{dt}$$

8.6. Based on the information given in the problem, we have:

$$N = 1$$

$$\varphi_1 = -0.1 \ wb$$

$$\varphi_2 = 0.9 \ wb$$

$$\Delta t = 0.125 \ s$$

As we know, the instantaneous voltage induced ($e(t)$) across a coil, inductor, or solenoid with N turns placed in a magnetic field with the magnetic flux of $\varphi(t)$ can be calculated as follows.

$$e(t) = -N\frac{d\varphi(t)}{dt}$$

Moreover, the average voltage induced ($\overline{e}$) across the coil can be calculated as follows.

$$\bar{e} = -N\frac{\Delta\varphi}{\Delta t} = -N\frac{\varphi_2 - \varphi_1}{t_2 - t_1}$$

Therefore:

$$\bar{e} = -1 \times \frac{0.9 - (-0.1)}{0.125}$$

$$\Rightarrow \bar{e} = -8\ V$$

Choice (1) is the answer.

8.7. As we know, the instantaneous voltage induced ($e(t)$) across a coil, inductor, or solenoid with N turns placed in a magnetic field with the magnetic flux of $\varphi(t)$ can be calculated as follows.

$$e(t) = -N\frac{d\varphi(t)}{dt}$$

Moreover, the average voltage induced ($\bar{e}$) across the coil can be calculated as follows.

$$\bar{e} = -N\frac{\Delta\varphi}{\Delta t} = -N\frac{\varphi_2 - \varphi_1}{t_2 - t_1}$$

Thus, for the first part of the magnetic flux-time curve, we can write:

$$e(t) = -\frac{d}{dt}(t^2) = -2t\ V$$

For the second part of the magnetic flux-time curve, we can write:

$$e(t) = -\frac{d}{dt}(1) = 0\ V$$

For the third part of the magnetic flux-time curve, we can write:

$$e(t) = -\frac{d}{dt}(-t + 3) = 1$$

Therefore, the induced voltage-time curve is like the one illustrated in Fig. 8.2b. Choice (4) is the answer.

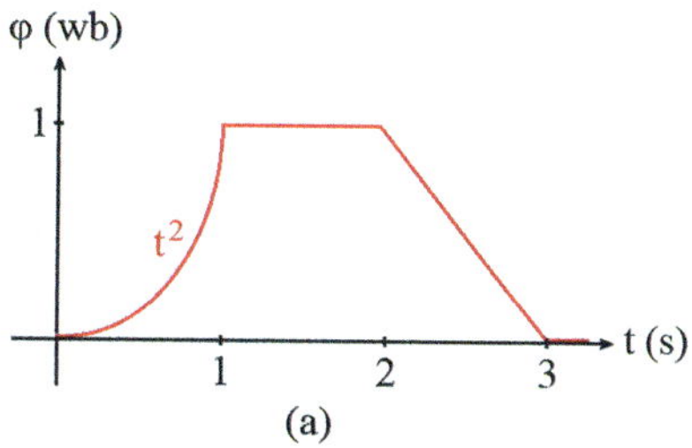

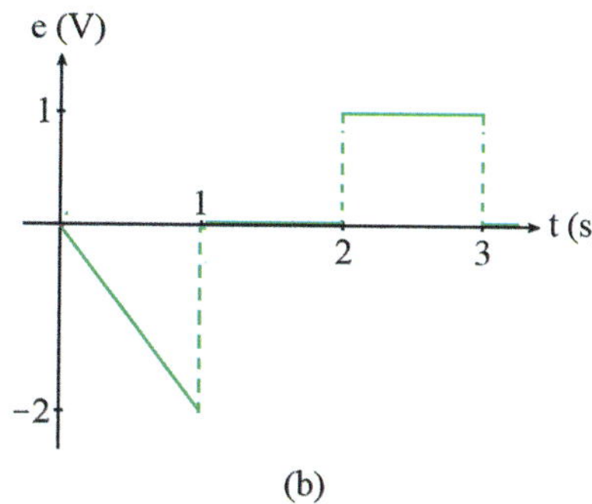

Fig. 8.2 (**a**) The magnetic flux-time curve. (**b**) The induced voltage-time curve

Notes

In this problem, the relations below have been used.

$$\frac{d}{dt}t^n = nt^{n-1}$$

$$\frac{d}{dt}a = 0$$

8.3 Voltage Induced in a Wire

8.8. Based on the information given in the problem, we have:

$$l = 2\ m$$

$$v = 20\frac{m}{s}$$

$$B = 1\ T$$

$$\theta = 90^\circ$$

The voltage induced in a straight wire with the length of l that moves at velocity $\vec{v}$ in an environment with the magnetic field of $\vec{B}$ can be calculated as follows.

$$e = l\vec{v} \times \vec{B} = Blv \sin\theta$$

where, θ is the angle between the $\vec{v}$ and $\vec{B}$.

Thus, for this problem, we have:

$$e = 1 \times 2 \times 20 \sin 90\,^\circ$$

$$\Rightarrow e = 40\ V$$

Choice (3) is the answer.

8.4 Inductance and Equivalent Inductance of a Circuit

8.10. Based on the information given in the problem, we have:

$$I_1 = 10\ A$$

$$I_2 = 0\ A$$

$$\Delta t = 2\ s$$

$$\overline{e} = 5\ V$$

If the current of a coil, inductor, or solenoid with the inductance of L changes with respect to time, an instantaneous voltage will be induced (e) across it that can be calculated as follows.

$$e(t) = -L\frac{dI}{dt}$$

Moreover, the average voltage induced ($\overline{e}$) across a coil, inductor, or solenoid can be calculated as follows.

$$\overline{e} = -L\frac{\Delta I}{\Delta t} = -L\frac{I_2 - I_1}{t_2 - t_1}$$

Therefore:

$$5 = -L\frac{0 - 10}{2}$$

$$\Rightarrow L = 1\ H$$

Choice (2) is the answer.

8.12. The equivalent inductance of the series inductors can be calculated as follows.

$$L_{eq} = 2 + 6 + 3$$

$$\Rightarrow L_{eq} = 11\ H$$

Choice (1) is the answer (Fig. 8.3).

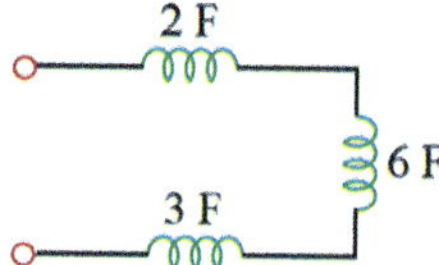

Fig. 8.3 The series connection of inductors

Notes

In this problem, the relations below have been used.

The equivalent inductance of n series inductors can be calculated as follows.

$$L_{eq} = L_1 + \cdots + L_n$$

If $L_1 = \cdots = L_n = L$, we have:

$$L_{eq} = nL$$

8.13. The equivalent inductance of the parallel inductors can be calculated as follows.

$$\frac{1}{L_{eq}} = \frac{1}{2} + \frac{1}{3} + \frac{1}{6} = \frac{3 + 2 + 1}{6} = \frac{6}{6}$$

$$\Rightarrow L_{eq} = 1\ H$$

Choice (2) is the answer (Fig. 8.4).

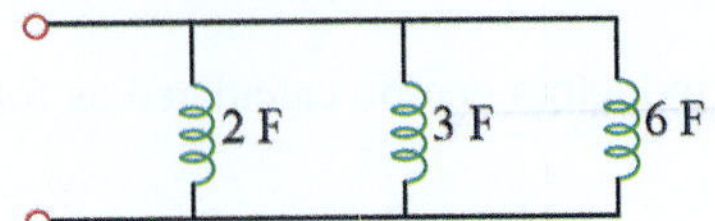

Fig. 8.4 The parallel connection of inductors

Notes

In this problem, the relations below have been used.

The equivalent inductance of n parallel inductors can be calculated as follows.

$$\frac{1}{L_{eq}} = \frac{1}{L_1} + \cdots + \frac{1}{L_n}$$

If $L_1 = \cdots = L_n = L$, we have:

$$L_{eq} = \frac{L}{n}$$

If $n = 2$, we have:

$$\frac{1}{L_{eq}} = \frac{1}{L_1} + \frac{1}{L_2} \Rightarrow L_{eq} = \frac{L_1 L_2}{L_1 + L_2}$$

8.14. In the circuit, the inductors are connected in series-parallel from. Therefore:

$$L_{eq} = 2 + 1||1$$

$$\Rightarrow L_{eq} = 2 + \frac{1 \times 1}{1 + 1}$$

$$\Rightarrow L_{eq} = 2.5\ H$$

Choice (4) is the answer (Fig. 8.5).

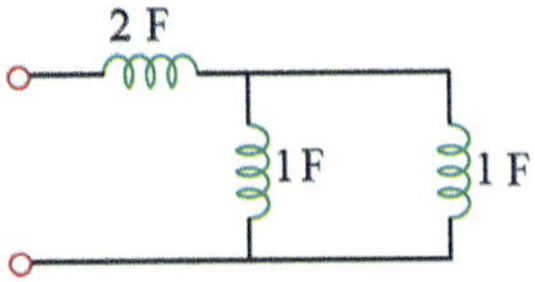

Fig. 8.5 The series-parallel connection of inductors

Notes

In this problem, the relations below have been used.

The equivalent inductance of n series inductors can be calculated as follows.

$$L_{eq} = L_1 + \cdots + L_n$$

If $L_1 = \cdots = L_n = L$, we have:

$$L_{eq} = nL$$

The equivalent inductance of n parallel inductors can be calculated as follows.

$$\frac{1}{L_{eq}} = \frac{1}{L_1} + \cdots + \frac{1}{L_n}$$

If $L_1 = \cdots = L_n = L$, we have:

$$L_{eq} = \frac{L}{n}$$

If $n = 2$, we have:

$$\frac{1}{L_{eq}} = \frac{1}{L_1} + \frac{1}{L_2} \Rightarrow L_{eq} = \frac{L_1 L_2}{L_1 + L_2}$$

8.5 Magnetic Energy Stored in Inductors

8.16. Based on the information given in the problem, we have:

$$L = 2\,H$$

$$I = 3\,A$$

The magnetic energy stored in an inductor can be calculated as follows.

$$W = \frac{1}{2} L I^2$$

Therefore:

$$W = \frac{1}{2} \times 2 \times 3^2$$

$$\Rightarrow W = 9\,J$$

Choice (2) is the answer.

References

1. Rahmani-Andebili, M., General Physics I - Practice Problems, Methods, and Solutions, Springer Nature, 2025.
2. Rahmani-Andebili, M., Calculus III – Practice Problems, Methods, and Solutions, Springer Nature, 2023.
3. Rahmani-Andebili, M., Calculus II – Practice Problems, Methods, and Solutions, Springer Nature, 2023.
4. Rahmani-Andebili, M., Calculus I (2nd Ed.) – Practice Problems, Methods, and Solutions, Springer Nature, 2023.
5. Rahmani-Andebili, M., Precalculus (2nd Ed.) – Practice Problems, Methods, and Solutions, Springer Nature, 2024.

9 Fluid Dynamics: Part A

Abstract

In this chapter, the basic and advanced problems of Fluid Dynamics are studied. The subjects include Density, Pressure in Fluids, Archimedes' Principle, and Continuity Principle of Fluids. Herein, different types of problems and exercises are presented that are categorized as follows.

- ***Problems with detailed solution***: They have been designed to teach students the subjects in detail. Moreover, they have been categorized in different levels based on their difficulty levels (easy, normal, and hard) and calculation amounts (small, normal, and large).
- ***Partially solved exercises***: They have been designed to encourage students to practice problems while guiding them through the problem-solving procedure and hinting the required formulas.
- ***Exercises with final answer***: They have been designed to encourage students to practice more by themselves while hinting them by the final answer as well as to help instructors to give tests or quizzes.

9.1 Density

Problem

9.1. The density of mercury at $20\ ^\circ C$ is $13.545 \times 10^3\ kg/m^3$. Express its density in gr/cm^3 [1–5].

Difficulty level ● Easy ○ Normal ○ Hard

Calculation amount ● Small ○ Normal ○ Large

1) $13.545 \times 10^6\ gr/cm^3$
2) $13.545 \times 10^3\ gr/cm^3$
3) $13.545 \times 10^{-3}\ gr/cm^3$
4) $13.545\ gr/cm^3$

Partially Solved Exercise

9.2. The density of gallium at $20\ ^\circ C$ is $5.907\ gr/cm^3$. Express its density in kg/m^3.

Solution

Based on the information given in the problem, we have:

$$\rho = 5.907\ gr/cm^3$$

As we know, $1\ gr = 10^{-3}\ kg$ and $1\ cm^3 = 10^{-6}\ m^3$. Thus:

M. Rahmani-Andebili, *General Physics II*, https://doi.org/10.1007/978-3-031-92866-6_9

$$\rho = 5.907 \times \frac{(\qquad kg)}{(\qquad m^3)} = 5.907 \times (\qquad) \; kg/cm^3$$

$$\Rightarrow \rho = 5907 \; kg/m^3$$

9.2 Pressure in Fluids

Problem

9.3. Calculate the pressure at 100-m depth of a sea resulted only from water. Herein, assume that the density of water is 1000 kg/m^3 and the gravitational acceleration is about 10 m/s^2. Herein, assume that 1 $atm = 10^5$ Pa.

Difficulty level ● Easy ○ Normal ○ Hard

Calculation amount ● Small ○ Normal ○ Large

1) 10 atm
2) 1 atm
3) 100 atm
4) 10^6 atm

Partially Solved Exercise

9.4. Calculate the total pressure at 100-m depth of a sea. Herein, assume that the density of water is 1000 kg/m^3, the gravitational acceleration is about 10 m/s^2, 1 $atm = 10^5$ Pa, and the air pressure at the surface of the sea is about 1 atm or 10^5 Pa.

Solution

Based on the information given in the problem, we have:

$$h = 100 \; m$$

$$\rho = 1000 \; kg/m^3$$

$$g = 10 \; m/s^2$$

$$1 \; atm = 10^5 \; Pa$$

$$P_0 = 10^5 \; Pa$$

The total pressure in a sea can be calculated as follows.

$$P_1 = P_0 + \rho g h$$

where, P_0 is the air pressure at the surface of the sea.

Hence:

$$P_1 = (\qquad) + (\qquad) \times (\qquad) \times (\qquad)$$

$$\Rightarrow P = (\qquad) \; Pa = 11 \; atm$$

Problem

9.5. An air bubble with the volume of 2 mm^3 is created at the bottom of a lake with the depth of 5 m. Calculate the volume of the bubble when it reaches the surface of the lake. Herein, assume that the temperature of the lake is constant everywhere, the air pressure at the surface of the lake is about 1 atm or 10^5 Pa, the density of air is 1 kg/m^3, and the gravitational acceleration is about 10 m/s^2.

Difficulty level ○ Easy ○ Normal ● Hard
Calculation amount ○ Small ● Normal ○ Large

1) 2 mm^3
2) 2.0005 mm^3
3) 2.001 mm^3
4) 2.01 mm^3

Partially Solved Exercise

9.6. If the temperature and volume of an ideal gas increases about 10%, how does its pressure change?

Solution

Based on the information given in the problem, we have:

$$V_2 = 1.1V_1$$

$$T_2 = 1.1T_1$$

As we know, for an ideal gas, we have:

$$\frac{P_1V_1}{T_1} = \frac{P_2V_2}{T_2}$$

Therefore:

$$\frac{P_1V_1}{T_1} = \frac{P_2(\quad)}{(\quad)}$$

$$\Rightarrow P_1 = (\quad)P_2$$

$$\Rightarrow P_2 = P_1$$

As can be seen, the pressure of the ideal gas does not change.

Problem

9.7. A barometer shows 350 kPa as the relative pressure of a tank; however, a local barometer shows 75 $cm\ Hg$. Calculate the absolute pressure of the tank by assuming that $g = 10\ \frac{m}{s^2}$ and $\rho_{Hg} = 13600\ \frac{kg}{m^3}$.

Difficulty level ○ Easy ● Normal ○ Hard
Calculation amount ● Small ○ Normal ○ Large

1) 248 Pa
2) 102 Pa
3) 452 Pa
4) 350 Pa

9.3 Archimedes' Principle

Problem

9.8. In an experiment, 60% of the volume of a piece of wood is immersed in water. Moreover, in the second experiment, 85% of the volume of the wood is immersed in a specific oil. Calculate the density of the oil. Herein, assume that the density of water is 1000 kg/m^3.

Difficulty level ○ Easy ○ Normal ● Hard
Calculation amount ○ Small ● Normal ○ Large

1) $706\,\frac{kg}{m^3}$
2) $600\,\frac{kg}{m^3}$
3) $510\,\frac{kg}{m^3}$
4) $1666\,\frac{kg}{m^3}$

Partially Solved Exercise

9.9. The density of a piece of a specific wood is 700 kg/m^3. Calculate the percentage of the volume of the wood that will immers in water. Herein, assume that the density of water is 1000 kg/m^3.

Solution

By using Archimedes' principle, we have:

$$\rho_{wood} V_{wood} = \rho_{water} V_{water}$$

Herein, ρ_{wood}, V_{wood}, ρ_{water}, and V_{water} are the density of wood (kg/m^3), volume of wood (m^3), density of water (kg/m^3), and volume of water that will spill outside (m^3) which is equal to the volume of wood that will immerse in water (m^3).

$$\Rightarrow (\quad\quad) V_{wood} = (\quad\quad) \times V_{water}$$

$$\Rightarrow \frac{V_{water}}{V_{wood}} = \frac{(\quad\quad)}{(\quad\quad)} = 0.7$$

Therefore, the percentage of the volume of the wood that will immers in water is as follows.

$$\frac{V_{water}}{V_{wood}} \times 100 = 70\%$$

9.4 Continuity Principle of Fluids

Problem

9.10. Two small rivers come together and make a large river with the width and velocity of 10 m and 3 m/s, respectively. The width, depth, and velocity of the first river are 8 m, 3 m, and 2 m/s, respectively. In addition, the width, depth, and velocity of the second river are 6 m, 3 m, and 3 m/s, respectively. Calculate the depth of the large river.

Difficulty level ○ Easy ● Normal ○ Hard
Calculation amount ○ Small ● Normal ○ Large

1) 3.4 m
2) 2.3 m
3) 4.3 m
4) 3.2 m

References

1. Rahmani-Andebili, M., General Physics I – Practice Problems, Methods, and Solutions, Springer Nature, 2025.
2. Rahmani-Andebili, M., Calculus III – Practice Problems, Methods, and Solutions, Springer Nature, 2023.
3. Rahmani-Andebili, M., Calculus II – Practice Problems, Methods, and Solutions, Springer Nature, 2023.
4. Rahmani-Andebili, M., Calculus I (2nd Ed.) – Practice Problems, Methods, and Solutions, Springer Nature, 2023.
5. Rahmani-Andebili, M., Precalculus (2nd Ed.) – Practice Problems, Methods, and Solutions, Springer Nature, 2024.

Fluid Dynamics: Part B

10

Abstract

In this chapter, the problems of the ninth chapter are fully solved, in detail, step-by-step, and with different methods.

10.1 Density

10.1. Based on the information given in the problem, we have [1–5]:

$$\rho = 13.545 \times 10^3\ kg/m^3$$

As we know, $1\ kg = 1000\ gr$ and $1\ m^3 = 10^6\ cm^3$. Thus:

$$\rho = 13.545 \times 10^3 \times \frac{1000\ gr}{10^6\ cm^3} = 13.545 \times 10^3 \times 10^{-3}\ gr/cm^3$$

$$\Rightarrow \rho = 13.545\ gr/cm^3$$

Choice (4) is the answer.

10.2 Pressure in Fluids

10.3. Based on the information given in the problem, we have:

$$h = 100\ m$$

$$\rho = 1000\ kg/m^3$$

$$g = 10\ m/s^2$$

$$1\ atm = 10^5\ Pa$$

M. Rahmani-Andebili, *General Physics II*, https://doi.org/10.1007/978-3-031-92866-6_10

The pressure in a sea resulted only from water can be calculated as follows.

$$P = \rho g h$$

Hence:

$$P = 1000 \times 10 \times 100$$

$$\Rightarrow P = 10^6\ Pa = 10\ atm$$

Choice (1) is the answer.

10.5. Based on the information given in the problem, we have:

$$V_1 = 2\ mm^3$$

$$h = 5\ m$$

$$T_1 = T_2$$

$$P_0 = 1\ atm = 10^5\ Pa$$

$$\rho = 1\ kg/m^3$$

$$g = 10\ m/s^2$$

As we know, for an ideal gas, we have:

$$\frac{P_1 V_1}{T_1} = \frac{P_2 V_2}{T_2}$$

Moreover, the total pressure at the bottom of a lake can be calculated as follows.

$$P_1 = P_0 + \rho g h$$

where, P_0 is the air pressure at the surface of the lake.

Therefore:

$$\frac{(P_0 + \rho g h) V_1}{T_1} = \frac{P_0 V_2}{T_1}$$

$$\Rightarrow \left(10^5 + 1 \times 10 \times 5\right) \times 2 \times 10^{-9} = 10^5 V_2$$

$$\Rightarrow 1.0005 \times 10^5 \times 2 \times 10^{-9} = 10^5 V_2$$

$$\Rightarrow 2.001 \times 10^{-4} = 10^5 V_2$$

$$\Rightarrow V_2 = 2.001 \times 10^{-9}\ m^3$$

$$\Rightarrow V_2 = 2.001\ mm^3$$

Choice (3) is the answer.

Notes
In this problem, the relation below has been used.

$$1\ m^3 = 10^9\ mm^3$$

10.7. Based on the information given in the problem, we have:

$$P_{relative} = 350\ kPa$$

$$P_0 = 75\ cm\ Hg$$

$$g = 10\ \frac{m}{s^2}$$

$$\rho_{Hg} = 13600\ \frac{kg}{m^3}$$

The air pressure in the area can be calculated in Pascal as follows.

$$P_0 = \rho g h = 13600 \times 10 \times 0.75 = 102\ kPa$$

The absolute pressure of the tank can be calculated as follows.

$$P_{absolute} = P_0 + P_{relative}$$

$$\Rightarrow P_{absolute} = 102 + 350$$

$$\Rightarrow P_{absolute} = 452\ kPa$$

Choice (3) is the answer.

10.3 Archimedes' Principle

10.8. Based on Archimedes' principle, for the first experiment, we have:

$$\rho_{wood} V_{wood} = \rho_{water} V_{water}$$

Herein, ρ_{wood}, V_{wood}, ρ_{water}, and V_{water} are the density of wood (kg/m^3), volume of wood (m^3), density of water (kg/m^3), and volume of water that will spill outside (m^3) which is equal to the volume of wood that will immerse in water (m^3).

$$\rho_{wood} V_{wood} = 1000 \times 0.6 V_{wood}$$

$$\Rightarrow \rho_{wood} = 600 \frac{kg}{m^3}$$

Applying Archimedes' principle for the second experiment:

$$\Rightarrow \rho_{wood} V_{wood} = \rho_{oil} V_{oil}$$

$$\Rightarrow 600 V_{wood} = \rho_{oil} \times 0.85 V_{wood}$$

$$\Rightarrow \rho_{oil} \approx 706 \frac{kg}{m^3}$$

Choice (1) is the answer.

10.4 Continuity Principle of Fluids

10.10. Based on the information given in the problem, we have:

$$w = 10\ m$$

$$v = 3\ m/s$$

$$w_1 = 8\ m$$

$$d_1 = 3\ m$$

$$v_1 = 2\ m/s$$

$$w_2 = 6\ m$$

$$d_2 = 3\ m$$

$$v_2 = 3\ m/s$$

Based on continuity principle of fluids, we can write:

$$A_1 v_1 + A_2 v_2 = A v$$

Herein, A and v stand for the surface area of the fluid and its velocity, respectively. Therefore:

$$w_1 d_1 v_1 + w_2 d_2 v_2 = w d v$$

$$8 \times 3 \times 2 + 6 \times 3 \times 3 = 10 \times d \times 3$$

$$\Rightarrow d = \frac{102}{30} \Rightarrow d = 3.4\ m$$

Choice (1) is the answer.

References

1. Rahmani-Andebili, M., General Physics I – Practice Problems, Methods, and Solutions, Springer Nature, 2025.
2. Rahmani-Andebili, M., Calculus III – Practice Problems, Methods, and Solutions, Springer Nature, 2023.
3. Rahmani-Andebili, M., Calculus II – Practice Problems, Methods, and Solutions, Springer Nature, 2023.
4. Rahmani-Andebili, M., Calculus I (2nd Ed.) – Practice Problems, Methods, and Solutions, Springer Nature, 2023.
5. Rahmani-Andebili, M., Precalculus (2nd Ed.) – Practice Problems, Methods, and Solutions, Springer Nature, 2024.

11 Thermodynamics: Part A

Abstract

In this chapter, the basic and advanced problems of Thermodynamics are studied. The subjects include Temperature and Thermometers, Thermal Expansion, Heat Transfer, Laws of Thermodynamics, Entropy and Change of Entropy, and Carnot Machine. Herein, different types of problems and exercises are presented that are categorized as follows.

- ***Problems with detailed solution***: They have been designed to teach students the subjects in detail. Moreover, they have been categorized in different levels based on their difficulty levels (easy, normal, and hard) and calculation amounts (small, normal, and large).
- ***Partially solved exercises***: They have been designed to encourage students to practice problems while guiding them through the problem-solving procedure and hinting the required formulas.
- ***Exercises with final answer***: They have been designed to encourage students to practice more by themselves while hinting them by the final answer as well as to help instructors to give tests or quizzes.

11.1 Temperature and Thermometers

Problem

11.1. At what temperature, the Celsius and Fahrenheit scales fit each other [1–5].

Difficulty level ● Easy ○ Normal ○ Hard

Calculation amount ● Small ○ Normal ○ Large

1) $-40\ °F$
2) $32\ °F$
3) $40\ °F$
4) $104\ °F$

Partially Solved Exercise

11.2. The temperature of a room is $25\ °C$. Express it in Fahrenheit and Kelvin scales.

Solution

The relation between Celsius and Fahrenheit scales is as follows.

$$\theta_F = 1.8\theta_C + 32$$

Moreover, the relation between Celsius and Kelvin scales is as follows.

M. Rahmani-Andebili, *General Physics II*, https://doi.org/10.1007/978-3-031-92866-6_11

$$\theta_K = \theta_C + 273$$

Therefore:

$$\theta_F = (\quad) \times (\quad) + (\quad) = 70\,°F$$

$$\theta_K = (\quad) + (\quad) = 298\,°K$$

11.2 Thermal Expansion

Problem

11.3. At 25 °C temperature, the length of two rods of different type is 3 *cm* and 2.9 *cm*, respectively. Determine the temperature in which their lengths are equal. The linear thermal expansion coefficient of the rods is $\alpha_a = 11 \times 10^{-5} \frac{1}{°C}$ and $\alpha_b = 20 \times 10^{-5} \frac{1}{°C}$, respectively.

Difficulty level ○ Easy ○ Normal ● Hard

Calculation amount ○ Small ○ Normal ● Large

1) 425 °*K*
2) 400 °*C*
3) 400 °*K*
4) 425 °*C*

Partially Solved Exercise

11.4. A 150 *cm* meterstick of a specific type is designed to measure accurate lengths at room temperature (25 °*C*). Calculate the thermal expansion of the meterstick if it is used outside at 45 °*C* and its linear thermal expansion coefficient is $\alpha = 11 \times 10^{-5} \frac{1}{°C}$.

Solution

Based on the information given in the problem, we have:

$$T_1 = 25\,°C$$

$$l_1 = 150\ cm$$

$$\alpha = 11 \times 10^{-5} \frac{1}{°C}$$

As we know, the relation between the length and temperature of a metal is as follows.

$$l_2 = l_1(1 + \alpha(T_2 - T_1))$$

$$\Rightarrow \Delta l = l_1 \alpha \Delta T$$

Herein, l_2, l_1, α, T_2, and T_1 are the final length (m), primary length (m), linear thermal expansion coefficient $\left(\frac{1}{°C}\right)$, final temperature (°*C*), and primary temperature (°*C*) of the metal, respectively.

Therefore, we can write:

$$\Delta l = (\qquad) \times (\qquad) \times ((\qquad) - (\qquad))$$

$$\Rightarrow \Delta l = 0.33\ cm$$

11.3 Heat Transfer

Problem

11.5. An object falls from a 50-m height. Calculate the thermal capacity of the object if its temperature rises about 20 °C. Herein, assume that $g = 10\ m/s^2$ and there is no energy loss.

Difficulty level ○ Easy ● Normal ○ Hard

Calculation amount ○ Small ● Normal ○ Large

1) 15 J/°C
2) 25 J/°C
3) 10 J/°C
4) 5 J/°C

Problem

11.6. The specific thermal capacity of two objects are $c_A = 0.5 \frac{J}{kg\,°C}$ and $c_B = 0.4 \frac{J}{kg\,°C}$. What mass of object B is equivalent to 80 gr of object A from thermal point of view?

Difficulty level ○ Easy ● Normal ○ Hard

Calculation amount ○ Small ● Normal ○ Large

1) 64 gr
2) 80 gr
3) 90 gr
4) 100 gr

11.4 Laws of Thermodynamics

Problem

11.7. Which one of the paths illustrated in Fig. 11.1 shows the maximum work done on a gas.

Difficulty level ● Easy ○ Normal ○ Hard

Calculation amount ● Small ○ Normal ○ Large

1) D
2) C
3) B
4) A

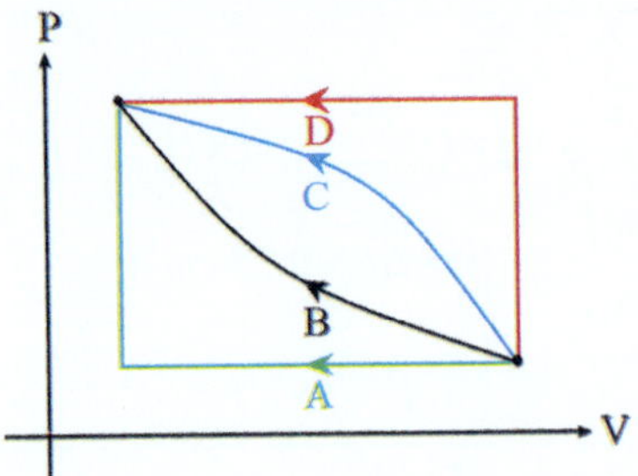

Fig. 11.1 The works done on a gas through different passes

Problem

11.8. Which of the following gasses has the largest value of $C_p - C_v$?

Difficulty level ● Easy ○ Normal ○ Hard

Calculation amount ● Small ○ Normal ○ Large

1) Monatomic gases.
2) Diatomic gases.
3) Polyatomic gases.
4) The value of $C_p - C_v$ is constant for any type of gas.

Problem

11.9. The temperature of an ideal gas is tripled from its initial temperature of 127 °*C*. Calculate the value of the following term in which v_{rms} is the root-mean-square velocity of the gas.

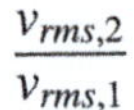

$$\frac{v_{rms,2}}{v_{rms,1}}$$

Difficulty level ○ Easy ● Normal ○ Hard

Calculation amount ○ Small ● Normal ○ Large

1) 1.635
2) $\sqrt{3}$
3) 1.28
4) 3

Exercise

11.10. The initial temperature of an ideal gas is about 27 °*C*. At what temperature its root-mean-square velocity (v_{rms}) will double?

Final Answer

$T_2 = 927\ °C$

Problem

11.11. A container includes some ideal gas. Which of the following choices is correct about the value of $\frac{m_2}{m_1}$ if the temperature of the gas increases from 77 °C to 177 °C at constant pressure. Herein, m is the mass of the gas.

Difficulty level ○ Easy ○ Normal ● Hard
Calculation amount ○ Small ● Normal ○ Large

1) It does not change.
2) $\frac{9}{7}$
3) $\frac{7}{9}$
4) $\frac{77}{177}$

Problem

11.12. Which one of the following choices is correct about the internal energy of an ideal gas.

Difficulty level ● Easy ○ Normal ○ Hard
Calculation amount ● Small ○ Normal ○ Large

1) The internal energy of an ideal gas depends only on its volume.
2) The internal energy of an ideal gas depends only on its pressure.
3) The internal energy of an ideal gas depends only on its temperature.
4) The internal energy of an ideal gas does not depend on the number of its molecules.

11.5 Entropy and Change of Entropy

Problem

11.13. Which of the following choices is correct about the entropy of the natural processes?

Difficulty level ● Easy ○ Normal ○ Hard
Calculation amount ● Small ○ Normal ○ Large

1) The entropy of natural processes does not change.
2) The entropy of natural processes decreases.
3) The entropy of natural processes increases.
4) The entropy of natural processes is not measurable.

Problem

11.14. In which one of the following processes, the change of entropy is zero?

Difficulty level ● Easy ○ Normal ○ Hard
Calculation amount ● Small ○ Normal ○ Large

1) Free expansion process
2) Isothermal processes (constant-temperature process)
3) Isochoric process (constant-volume process)
4) Nonreversible adiabatic process

Problem

11.15. The temperature of a 20 gr solid with the specific thermal capacity of 225 $\frac{J}{kg\,°C}$ increases from 27 $°C$ to 177 $°C$ in a process. Calculate the change in the entropy of the solid.

Difficulty level ○ Easy ● Normal ○ Hard

Calculation amount ○ Small ● Normal ○ Large

1) $4.5\ln\left(\frac{177}{27}\right)\ \frac{J}{°C}$
2) $4.5\ln(1.5)\ \frac{J}{°C}$
3) $4.5\ \frac{J}{°C}$
4) $4500\ \frac{J}{°C}$

Problem

11.16. In a process, 4 kg water at 20 $°C$ and 1 kg water at 60 $°C$ are mixed. Calculate the change in the entropy of the system. Herein, assume that $c_{water} = 4200\ \frac{J}{kg\,°C}$ and there is no energy loss.

Difficulty level ○ Easy ○ Normal ● Hard

Calculation amount ○ Small ○ Normal ● Large

1) $40.4\ \frac{J}{°C}$
2) $28.3\ \frac{J}{°C}$
3) $-8.3\ \frac{J}{°C}$
4) $-16.7\ \frac{J}{°C}$

11.6 Carnot Machine

Problem

11.17. Calculate the efficiency of a Carnot machine that its low and high temperatures are 27 $°C$ and 227 $°C$.

Difficulty level ○ Easy ● Normal ○ Hard

Calculation amount ● Small ○ Normal ○ Large

1) 25%
2) 40%
3) 66.6%
4) 88.1%

Partially Solved Exercise

11.18. The efficiency of a Carnot machine is about 22%. Calculate the low and high temperatures of the machine if their difference is 75 $°C$.

Solution

Based on the information given in the problem, we have:

$$\eta = 0.22 \tag{11.1}$$

$$T_H - T_L = 75\,°C \tag{11.2}$$

As we know, the efficiency of a Carnot machine can be calculated by each of the following relations.

$$\eta = 1 - \frac{T_L}{T_H} \tag{11.3}$$

$$\eta = \frac{W}{Q_H} = 1 - \frac{Q_L}{Q_H} \tag{11.4}$$

Therefore:

$$(\quad) = 1 - \frac{T_L}{T_H} \Rightarrow \frac{T_L}{T_H} = (\quad) \tag{11.5}$$

Solving (11.2) and (11.5):

$$T_H - (\quad)T_H = (\quad)$$

$$\Rightarrow T_H = 340.9\,°C \tag{11.6}$$

Solving (11.2) and (11.6):

$$(\quad) - T_L = (\quad)$$

$$\Rightarrow T_L = 265.9\,°C$$

Problem

11.19. In a Carnot machine for 100 *kJ* output work, 150 *kJ* energy enters the cold source. Calculate the high temperature if the low temperature of the machine is 27 °*C*.

Difficulty level ○ Easy ● Normal ○ Hard

Calculation amount ○ Small ● Normal ○ Large

1) 500 °C
2) 125 °C
3) 250 °C
4) 227 °C

Problem

11.20. In Fig. 11.2, we have $Q_{CA} = -5\ J$ and $Q_{AB} = 80\ J$. Calculate the efficiency of the cycle.

Difficulty level ○ Easy ● Normal ○ Hard

Calculation amount ○ Small ● Normal ○ Large

1) 40%
2) 50%
3) 60%
4) 80%

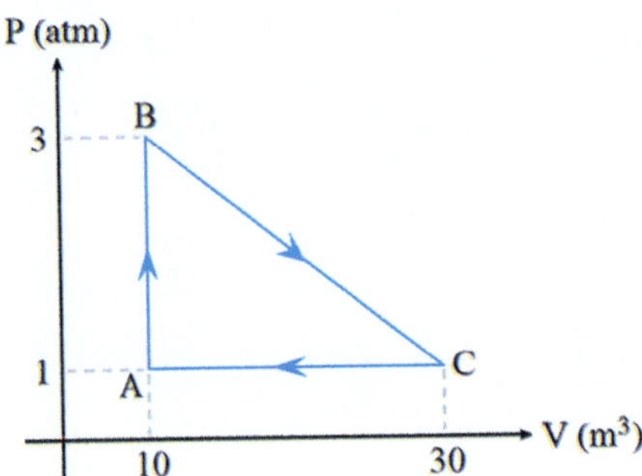

Fig. 11.2 The PV curve related to a thermal machine

Partially Solved Exercise

11.21. In Fig. 11.3, calculate the amount of heat absorbed by the thermal machine and the amount of work done by it.

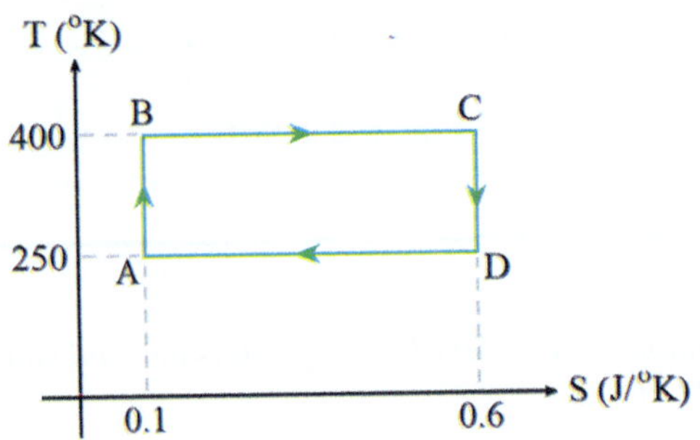

Fig. 11.3 The TS curve concerned with a thermal machine

Solution

As we know, the area of the cycle in a T-S graph is the work provided by the machine and the area under the BC line segment is the heat absorbed by the machine. Therefore:

$$W = ((\quad) - (\quad))((\quad) - (\quad)) = (\quad) \times (\quad)$$

$$\Rightarrow W = 75\ J$$

$$Q_H = (\quad)((\quad) - (\quad)) = (\quad) \times (\quad)$$

$$\Rightarrow Q_H = 200\ J$$

References

1. Rahmani-Andebili, M., General Physics I – Practice Problems, Methods, and Solutions, Springer Nature, 2025.
2. Rahmani-Andebili, M., Calculus III – Practice Problems, Methods, and Solutions, Springer Nature, 2023.
3. Rahmani-Andebili, M., Calculus II – Practice Problems, Methods, and Solutions, Springer Nature, 2023.
4. Rahmani-Andebili, M., Calculus I (2nd Ed.) – Practice Problems, Methods, and Solutions, Springer Nature, 2023.
5. Rahmani-Andebili, M., Precalculus (2nd Ed.) – Practice Problems, Methods, and Solutions, Springer Nature, 2024.

12 Thermodynamics: Part B

Abstract
In this chapter, the problems of the eleventh chapter are fully solved, in detail, step-by-step, and with different methods.

12.1 Temperature and Thermometers

12.1. The relation between Celsius and Fahrenheit scales is as follows [1–5].

$$\theta_F = 1.8\theta_C + 32 \quad (12.1)$$

On the other hand, we need to find the solution of the equation below.

$$\theta_F = \theta_C \quad (12.2)$$

Solving (12.1) and (12.2):

$$\theta_F = 1.8\theta_F + 32$$

$$\Rightarrow 0.8\theta_F = -32$$

$$\Rightarrow \theta_F = -40\,^\circ F$$

Choice (1) is the answer.

12.2 Thermal Expansion

12.3. Based on the information given in the problem, we have:

$$T_1 = 25\,^\circ C \quad (12.3)$$

$$l_{a,1} = 3\ cm \quad (12.4)$$

$$l_{b,1} = 2.9\ cm \quad (12.5)$$

M. Rahmani-Andebili, *General Physics II*, https://doi.org/10.1007/978-3-031-92866-6_12

$$l_{a,2} = l_{b,2} \tag{12.6}$$

$$\alpha_a = 11 \times 10^{-5} \frac{1}{^\circ C} \tag{12.7}$$

$$\alpha_b = 20 \times 10^{-5} \frac{1}{^\circ C} \tag{12.8}$$

As we know, the relation between the length and temperature of a metal is as follows.

$$l_2 = l_1(1 + \alpha(T_2 - T_1)) \tag{12.9}$$

$$\Rightarrow l_2 = l_1 + l_1 \alpha \Delta T \tag{12.10}$$

$$\Rightarrow \Delta l = l_1 \alpha \Delta T \tag{12.11}$$

Herein, l_2, l_1, α, T_2, and T_1 are the final length (m), primary length (m), linear thermal expansion coefficient $\left(\frac{1}{^\circ C}\right)$, final temperature ($^\circ C$), and primary temperature ($^\circ C$) of the metal, respectively.

Therefore, for the first one, we can write:

$$l_{a,2} = l_{a,1} + l_{a,1} \alpha_a \Delta T \tag{12.12}$$

$$\Rightarrow l_{a,2} = 0.03 + 0.03 \times 11 \times 10^{-5} \times \Delta T \tag{12.13}$$

Likewise, for the second one, we can write:

$$l_{b,2} = l_{b,1} + l_{b,1} \alpha_b \Delta T \tag{12.14}$$

$$\Rightarrow l_{b,2} = 0.029 + 0.029 \times 20 \times 10^{-5} \times \Delta T \tag{12.15}$$

Solving (12.16), (12.13), and (12.15):

$$0.03 + 0.03 \times 11 \times 10^{-5} \times \Delta T = 0.029 + 0.029 \times 20 \times 10^{-5} \times \Delta T$$

$$\Rightarrow \Delta T = \frac{0.03 - 0.029}{0.029 \times 20 \times 10^{-5} - 0.03 \times 11 \times 10^{-5}}$$

$$\Rightarrow \Delta T = \frac{0.001 \times 10^5}{0.029 \times 20 - 0.03 \times 11}$$

$$\Rightarrow \Delta T = \frac{100}{0.25} \Rightarrow \Delta T = 400$$

$$\Rightarrow T_2 - 25 = 400 \Rightarrow T_2 = 425\,^\circ C$$

Choice (4) is the answer.

12.3 Heat Transfer

12.5. Based on the information given in the problem, we have:

$$h = 50\ \text{m}$$

$$\Delta T = 20\,^\circ\text{C}$$

$$g = 10\ \text{m}/\text{s}^2$$

The primary mechanical energy of the object includes its gravitational potential energy. In other words:

$$E_1 = U_1 = mgh$$

As we know, the relation between the amount of heat (Q) that an object absorbs or releases and its temperature change (ΔT) is as follows.

$$Q = mc\Delta T$$

Thus, the final mechanical energy of the object includes its heat. Hence:

$$E_2 = Q$$

Based on the principle of conservation of mechanical energy, we have:

$$E_1 = E_2$$

$$\Rightarrow mgh = mc\Delta T$$

$$\Rightarrow gh = c\Delta T$$

$$\Rightarrow c = \frac{gh}{\Delta T} = \frac{10 \times 50}{20}$$

$$\Rightarrow c = 25\ J/\,^\circ C$$

Choice (2) is the answer.

12.6. Based on the information given in the problem, we have:

$$c_A = 0.5 \frac{J}{kg\,^\circ C}$$

$$c_B = 0.4 \frac{J}{kg\,^\circ C}$$

$$m_A = 80\ gr$$

To solve this problem, the thermal capacity of the masses must be equated as follows.

$$C_A = C_B$$

$$m_A c_A = m_B c_B$$

$$\Rightarrow 0.08 \times 0.5 = 0.4 m_B$$

$$\Rightarrow m_B = 0.1\ kg = 100\ gr$$

Choice (4) is the answer.

Notes

In this problem, the relation below has been used.

$$C = mc$$

12.4 Laws of Thermodynamics

12.7. The area under the P-V curve of a gas is equal to the amount of work done on the gas or by the gas. As can be noticed from Fig. 12.1, Path D has the largest amount of area. Hence, by Path D, the maximum amount of work is done on the gas. Choice (1) is the answer.

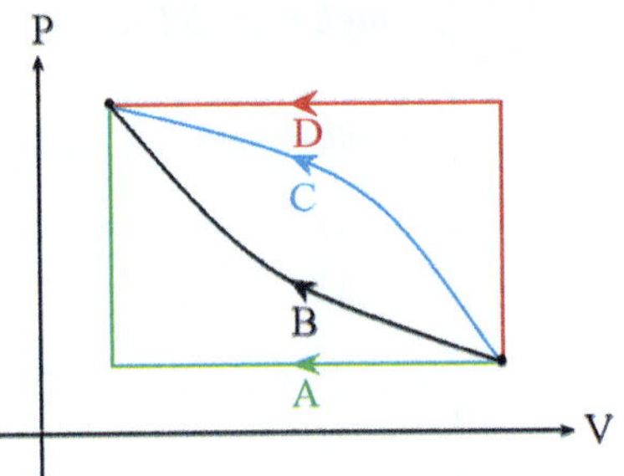

Fig. 12.1 The works done on a gas through different passes

12.8. The value of $C_p - C_v$ is constant and independent of the type of a gas. Choice (4) is the answer.

12.9. Based on the information given in the problem, we have:

$$T_1 = 127\,^\circ C$$

$$T_2 = 3T_1$$

The root-mean-square velocity of a gas can be calculated as follows.

$$v_{rms} = \sqrt{\frac{3RT}{M}}$$

where, R is the ideal gas constant $\left(\text{which is equal to } 8.314 \frac{J}{mol.^\circ K}\right)$, T is the temperature of the gas (in Kelvin), and M is the molar mass of the gas.

As can be learned from the problem, the only parameter that changes is the temperature of the gas. Thus:

$$\Rightarrow \frac{v_{rms,2}}{v_{rms,1}} = \sqrt{\frac{T_2}{T_1}}$$

$$\Rightarrow \frac{v_{rms,2}}{v_{rms,1}} = \sqrt{\frac{3 \times 127 + 273}{127 + 273}} = \sqrt{\frac{654}{400}} = \sqrt{1.635}$$

$$\Rightarrow \frac{v_{rms,2}}{v_{rms,1}} \approx 1.28$$

Choice (3) is the answer.

Notes

In this problem, the relation below has been used.

$$T_K = T_C + 273$$

12.11. Based on the information given in the problem, we have:

$$T_1 = 77^\circ C$$

$$T_2 = 177^\circ C$$

$$P_1 = P_2$$

As we know, the relation below is held for ideal gases.

$$PV = nRT$$

Thus:

$$\frac{P_1 V_1}{T_1} = \frac{P_2 V_2}{T_2}$$

$$\Rightarrow \frac{V_1}{77 + 273} = \frac{V_2}{177 + 273}$$

$$\Rightarrow \frac{V_2}{V_1} = \frac{450}{350}$$

$$\Rightarrow \frac{V_2}{V_1} = \frac{9}{7}$$

On the other hand, we know that:

$$\rho = \frac{m}{V}$$

Hence:

$$\frac{m_1}{V_1} = \frac{m_2}{V_2}$$

$$\Rightarrow \frac{m_2}{m_1} = \frac{V_2}{V_1}$$

$$\Rightarrow \frac{m_2}{m_1} = \frac{9}{7}$$

Choice (2) is the answer.

12.12. The internal energy of an ideal gas depends only on its temperature.

Choice (3) is the answer.

12.5 Entropy and Change of Entropy

12.13. The entropy of natural processes always increases.

Choice (3) is the answer.

12.14. In a nonreversible adiabatic process, the change of heat is zero ($\Delta Q = 0$). Therefore, the change of entropy (ΔS) is zero based on the relation below.

$$\Delta S = \frac{\int dQ}{T} = 0$$

Choice (4) is the answer.

12.15. Based on the information given in the problem, we have:

$$m = 20\ gr = 0.02\ kg$$

$$c = 225\ \frac{J}{kg\,^{\circ}C}$$

$$T_{initial} = 27\,^{\circ}C$$

$$T_{final} = 177\,^{\circ}C$$

The change in the entropy of a solid or liquid can be calculated as follows.

$$\Delta S = mc \ln \frac{T_{final}}{T_{initial}}$$

where, m, c, T_{final}, and $T_{initial}$ are the mass, specific thermal capacity, final temperature, and initial temperature, respectively. Herein, T_{final} and $T_{initial}$ must be presented in Kelvin.

$$\Rightarrow \Delta S = 0.02 \times 225 \times \ln \frac{273 + 177}{273 + 27}$$

$$\Rightarrow \Delta S = 4.5 \ln \frac{450}{300}$$

$$\Rightarrow \Delta S = 4.5 \ln 1.5\ \frac{J}{^{\circ}C}$$

Choice (2) is the answer.

Notes

In this problem, the relation below has been used.

$$T_K = T_C + 273$$

12.16. Based on the information given in the problem, we have:

$$m_1 = 4\ kg$$

$$T_{initial,1} = 20\,^{\circ}C$$

$$m_2 = 1\ kg$$

$$T_{initial,2} = 60\,^{\circ}C$$

$$c_{water} = 4200\ \frac{J}{kg\,^{\circ}C}$$

As we know, the change in the entropy of a solid or liquid can be calculated as follows.

$$\Delta S = mc \ln \frac{T_{final}}{T_{initial}}$$

where, m, c, T_{final}, and $T_{initial}$ are the mass, specific thermal capacity, final temperature, and initial temperature, respectively. Herein, T_{final} and $T_{initial}$ must be presented in Kelvin.

As can be seen, the final temperature of the mixture of water needs to be determined. Since there is no energy loss, we can write:

$$Q_1 = Q_2$$

$$m_1 c_1 \left(T_{final} - T_{initial,1}\right) = m_2 c_2 \left(T_{initial,2} - T_{final}\right)$$

$$\Rightarrow m_1 \left(T_{final} - T_{initial,1}\right) = m_2 \left(T_{initial,2} - T_{final}\right)$$

$$\Rightarrow 4 \times \left(T_{final} - (20 + 273)\right) = 1 \times \left((60 + 273) - T_{final}\right)$$

$$\Rightarrow 4\left(T_{final} - 293\right) = 333 - T_{final}$$

$$5T_{final} = 1505$$

$$T_{final} = 301\,^{\circ}K$$

Therefore, the change in the entropy of the system is as follows.

$$\Delta S = m_1 c_1 \ln \frac{T_{final}}{T_{initial,1}} + m_2 c_2 \ln \frac{T_{final}}{T_{initial,2}}$$

$$\Rightarrow \Delta S = 4 \times 4200 \times \ln \frac{301}{20 + 273} + 1 \times 4200 \times \ln \frac{301}{60 + 273}$$

$$\Rightarrow \Delta S = 28.2\ J/^{\circ}C$$

Choice (2) is the answer.

Notes

In this problem, the relations below have been used.

$$T_K = T_C + 273$$

The relation between the amount of heat (Q) that an object absorbs or releases and its temperature change (ΔT) is as follows. Herein, m and c are the mass and specific thermal capacity of the object, respectively.

$$Q = mc\left(T_{final} - T_{initial}\right)$$

12.6 Carnot Machine

12.17. Based on the information given in the problem, we have:

$$T_L = 27\,^\circ C$$

$$T_H = 227\,^\circ C$$

As we know, the efficiency of a Carnot machine can be calculated by each of the following relations.

$$\eta = 1 - \frac{T_L}{T_H}$$

$$\eta = \frac{W}{Q_H} = 1 - \frac{Q_L}{Q_H}$$

Thus:

$$\eta = 1 - \frac{27 + 273}{227 + 273} = 1 - \frac{300}{500}$$

$$\Rightarrow \eta = 0.4 = 40\%$$

Choice (2) is the answer.

Notes
In this problem, the relation below has been used.

$$T_K = T_C + 273$$

12.19. Based on the information given in the problem, we have:

$$W = 100\ kJ$$

$$Q_L = 150\ kJ$$

$$T_L = 27\,^\circ C$$

In a Carnot machine, the relations below are held.

$$Q_H = W + Q_L$$

$$\frac{T_L}{T_H} = \frac{Q_L}{Q_H}$$

Hence:

$$Q_H = 100 + 150 \Rightarrow Q_H = 250\ kJ$$

Therefore:

$$\frac{273+27}{T_H} = \frac{150}{250} \Rightarrow T_H = \frac{300 \times 250}{150}$$

$$\Rightarrow T_H = 500\,^{\circ}K$$

$$\Rightarrow T_H = 500 - 273 \Rightarrow T_H = 227\,^{\circ}C$$

Notes

In this problem, the relation below has been used.

$$T_K = T_C + 273$$

Choice (4) is the answer.

12.20. Based on the information given in the problem, we have:

$$Q_{CA} = -5\ J$$

$$Q_{AB} = 80\ J$$

As we know, the efficiency of a cycle can be calculated as follows.

$$\eta = \frac{W}{Q_H}$$

Moreover, we know that the area of the cycle in a P-V graph is the work provided by the machine and the area under the BC line segment is the heat absorbed by the machine. Therefore:

$$W = \frac{1}{2} \times 20 \times 2 = 20\ J$$

$$Q_H = \frac{1}{2}(3+1) \times 20 = 40\ J$$

$$\Rightarrow \eta = \frac{20}{40} = 0.5 \Rightarrow \eta = 50\%$$

Choice (2) is the answer (Fig. 12.2).

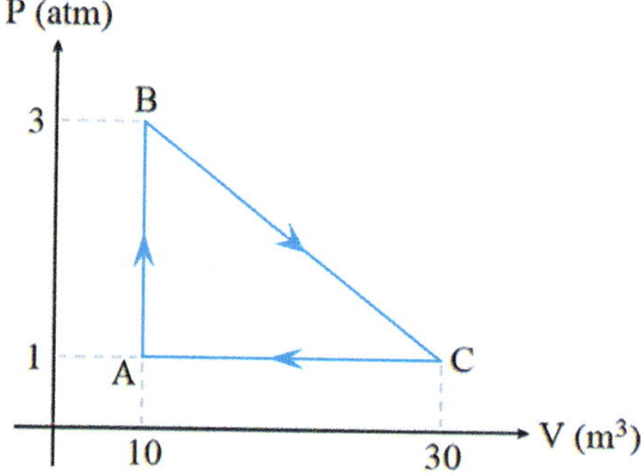

Fig. 12.2 The PV curve related to a thermal machine

References

1. Rahmani-Andebili, M., General Physics I – Practice Problems, Methods, and Solutions, Springer Nature, 2025.
2. Rahmani-Andebili, M., Calculus III – Practice Problems, Methods, and Solutions, Springer Nature, 2023.
3. Rahmani-Andebili, M., Calculus II – Practice Problems, Methods, and Solutions, Springer Nature, 2023.
4. Rahmani-Andebili, M., Calculus I (2nd Ed.) – Practice Problems, Methods, and Solutions, Springer Nature, 2023.
5. Rahmani-Andebili, M., Precalculus (2nd Ed.) – Practice Problems, Methods, and Solutions, Springer Nature, 2024.

13 Transverse and Longitudinal Waves: Part A

Abstract

In this chapter, the basic and advanced problems of Transverse and Longitudinal Waves are studied. The subjects include Transverse Waves, Longitudinal Waves, Sound Intensity, Sound Intensity Level, and Doppler Effect. Herein, different types of problems and exercises are presented that are categorized as follows.

- ***Problems with detailed solution***: They have been designed to teach students the subjects in detail. Moreover, they have been categorized in different levels based on their difficulty levels (easy, normal, and hard) and calculation amounts (small, normal, and large).
- ***Partially solved exercises***: They have been designed to encourage students to practice problems while guiding them through the problem-solving procedure and hinting the required formulas.
- ***Exercises with final answer***: They have been designed to encourage students to practice more by themselves while hinting them by the final answer as well as to help instructors to give tests or quizzes.

13.1 Transverse Waves

Problem

13.1. Which one of the points on the transverse wave, shown in Fig. 13.1, is in-phase with point A [1–5]?

Difficulty level ● Easy ○ Normal ○ Hard

Calculation amount ● Small ○ Normal ○ Large

1) B
2) C
3) D
4) E

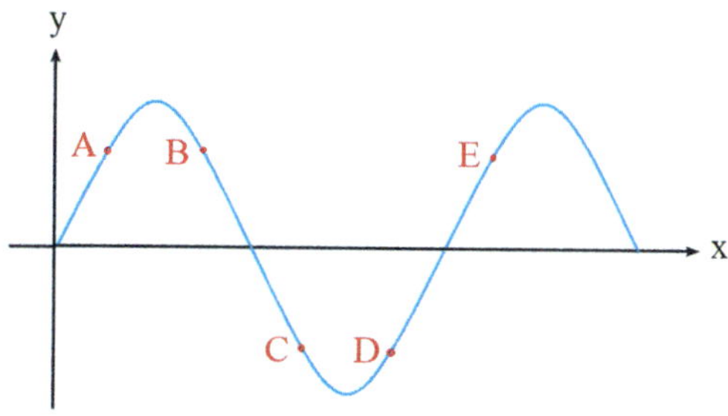

Fig. 13.1 A transverse wave

M. Rahmani-Andebili, *General Physics II*, https://doi.org/10.1007/978-3-031-92866-6_13

Problem

13.2. The transverse waves with the period and wavelength of 0.25 *s* and 10 *m* are propagating in the environment. Calculate their propagation velocity.

Difficulty level ● Easy ○ Normal ○ Hard

Calculation amount ● Small ○ Normal ○ Large

1) 10 *m/s*
2) 15 *m/s*
3) 20 *m/s*
4) 40 *m/s*

Exercise

13.3. The transverse waves with the frequency and velocity of 60 *Hz* and 300 *m/s* are propagating in the environment. Calculate their wavelength.

Final Answer

$\lambda = 5\ m$

Problem

13.4. A source propagates transverse waves with the frequency and wavelength of 60 *Hz* and 0.5 *m*. Calculate the time that will take for these waves to travel about 200 *m* distance.

Difficulty level ○ Easy ● Normal ○ Hard

Calculation amount ● Small ○ Normal ○ Large

1) 6.66 *s*
2) 66.66 *s*
3) 0.66 *s*
4) 60 *s*

Partially Solved Exercise

13.5. The transverse waves with the equation of $y(t) = A \sin 40\pi t$ are propagating at the velocity of 20 *m/s*. Calculate the distance between two consecutive in-phase points on the propagation direction.

Solution

Based on the information given in the problem, we have:

$$y(t) = A \sin 40\pi t \tag{13.1}$$

$$v = 20\ m/s \tag{13.2}$$

As we know, the distance between two consecutive in-phase points is the wavelength of a wave (λ). Hence, we need to calculate the value of λ.

From (13.1), the frequency of the wave can be extracted as follows.

$$\omega = 40\pi \Rightarrow 2\pi f = 40\pi \Rightarrow f = (\quad)\ Hz$$

As we know, the relation below exists between the velocity (m/s), frequency (Hz), wavelength (m), and period (s) of a transverse wave.

$$v = f\lambda = \frac{\lambda}{T}$$

Thus:

$$\lambda = \frac{v}{f} = \frac{(\quad)}{(\quad)}$$

$$\Rightarrow \lambda = 1\ m$$

Notes

In this problem, the relation below has been used.

$$\omega = 2\pi f$$

Herein, ω, f are the angular frequency (rad/s) and frequency (Hz), respectively.

The general form of a wave equation is as follows.

$$y(t) = A\sin(\omega t + \varphi_0)$$

Herein, A, ω, and φ_0 are the amplitude, angular frequency, and primary phase angle of the wave, respectively.

Problem

13.6. The ratio of propagation velocities of a transverse wave in water to air is four. Calculate the ratio of wavelength of the wave in water to air.

Difficulty level ○ Easy ● Normal ○ Hard
Calculation amount ● Small ○ Normal ○ Large

1) 1
2) 0.25
3) 4
4) 0

Exercise

13.7. The ratio of propagation velocities of a transverse wave in a medium to air is two. Calculate the ratio of the period of the wave in the medium to air.

Final Answer

$$\frac{T_{medium}}{T_{air}} = 0.5$$

Problem

13.8. The wave equation at a point is $u(x,t) = 0.1 \sin\left(\pi t - \frac{\pi}{3}\right)$. Calculate the phase change of the point during 0.05 s.

Difficulty level ○ Easy ● Normal ○ Hard
Calculation amount ● Small ○ Normal ○ Large

1) 9°
2) $0.05\pi°$
3) 0.05°
4) 18°

13.2 Longitudinal Waves

Problem

13.9. A two-meter, 0.2 kg string is tightened between two points. Calculate the propagation velocity of waves if the tension force of the string is 10 N.

Difficulty level ○ Easy ● Normal ○ Hard
Calculation amount ● Small ○ Normal ○ Large

1) 20 m/s
2) $\sqrt{20}$ m/s
3) $\sqrt{10}$ m/s
4) 10 m/s

Problem

13.10. Calculate the ratio of propagation velocities of waves if the intensity of propagation source is halved.

Difficulty level ● Easy ○ Normal ○ Hard
Calculation amount ● Small ○ Normal ○ Large

1) 0.5
2) 1
3) 2
4) 4

Problem

13.11. Calculate the propagation velocity of waves in a string with the density, cross-sectional area, and tension force of 2 gr/cm^3, 2 mm^2, and 100 N, respectively.

Difficulty level ○ Easy ● Normal ○ Hard
Calculation amount ● Small ○ Normal ○ Large

1) 0.1581 m/s
2) 1.581 m/s
3) 15.81 m/s
4) 158.1 m/s

Problem

13.12. In a guitar, the tension force of two strings is the same; however, the cross-sectional area of the first string is twice that of the second one. Calculate the ratio of propagation velocities of waves in the first string to the second one.

Difficulty level ○ Easy ○ Normal ● Hard

Calculation amount ○ Small ● Normal ○ Large

1) $\sqrt{2}$
2) $\frac{\sqrt{2}}{2}$
3) 2
4) $\frac{1}{2}$

Partially Solved Exercise

13.13. Calculate the ratio of propagation velocities of waves of a string if the length of string is halved but its tension force is quadrupled.

Solution

Based on the information given in the problem, we have:

$$l_2 = 0.5 l_1$$

$$F_2 = 4F_1$$

The propagation velocity of transverse waves in a string tightened between two points can be calculated by one of the following relations.

$$v = \sqrt{\frac{Fl}{m}}$$

$$v = \sqrt{\frac{F}{\rho A}}$$

Herein, F, l, m, ρ, A and are the tension force (N), length (m), mass (kg), density (kg/m^3), and surface area of the string (m^2), respectively.

Since the length of the string has been halved, its mass will be halved as well. In other words:

$$m_2 = (\quad) m_1$$

Hence:

$$\frac{v_2}{v_1} = \sqrt{\frac{F_2}{F_1} \times \frac{L_2}{L_1} \times \frac{m_1}{m_2}}$$

$$\Rightarrow \frac{v_2}{v_1} = \sqrt{(\quad) \times (\quad) \times (\quad)}$$

$$\Rightarrow \frac{v_2}{v_1} = 2$$

13.3 Sound Intensity and Sound Intensity Level

Problem

13.14. Which one of the following choices is correct about the frequency of sound in water and air.

Difficulty level • Easy ○ Normal ○ Hard

Calculation amount • Small ○ Normal ○ Large

1) The frequency of sound in water and air is the same.
2) The frequency of sound in water is larger than the one in air.
3) The frequency of sound in water is smaller than the one in air.
4) The frequency of sound depends on the temperature of the medium.

Problem

13.15. Which one of the following choices is correct about the speed of sound in water and air.

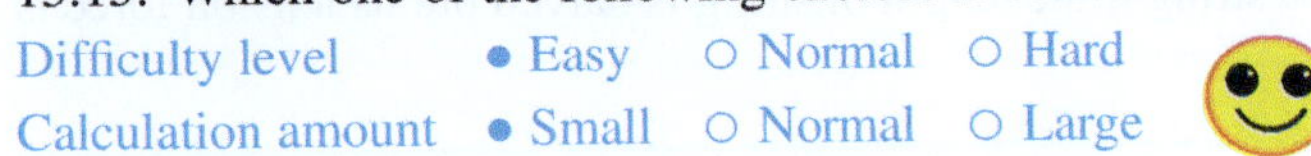

Difficulty level • Easy ○ Normal ○ Hard

Calculation amount • Small ○ Normal ○ Large

1) The speed of sound in water and air is the same.
2) The speed of sound in water is larger than the one in air.
3) The speed of sound in water is smaller than the one in air.
4) The speed of sound depends on the frequency of the sound source.

Problem

13.16. A sound source propagates the sound waves with 100 W power. Determine the distance where the sound intensity is about the threshold of pain. The sound intensity for threshold of pain is $I_{pain} = 1\ W/m^2$.

Difficulty level ○ Easy • Normal ○ Hard

Calculation amount • Small ○ Normal ○ Large

1) $r \approx 7.96\ m$
2) $r \approx 2.82\ m$
3) $r \approx 0.01\ m$
4) $r \approx 100\ m$

Problem

13.17. If the distance from a sound source is tripled, how does the ratio of sound intensities change?

Difficulty level ○ Easy • Normal ○ Hard

Calculation amount • Small ○ Normal ○ Large

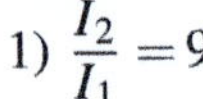

1) $\frac{I_2}{I_1} = 9$
2) $\frac{I_2}{I_1} = 3$
3) $\frac{I_2}{I_1} = \frac{1}{9}$
4) $\frac{I_2}{I_1} = \frac{1}{3}$

Partially Solved Exercise

13.18. If the amplitude of a sound source is tripled but its frequency and the distance from it are halved, how does the ratio of sound intensities change?

Solution

Based on the information given in the problem, we have:

$$A_2 = 3A_1$$

$$f_2 = \frac{1}{2}f_1$$

$$r_2 = \frac{1}{2}r_1$$

As we know, the sound intensity can be calculated as follows.

$$I = \frac{P}{4\pi r^2}$$

Herein, I, P, and r are the sound intensity (W/m^2), power of sound (W), and distance from the sound source (m), respectively.

Moreover, the sound power is proportional to the squared value of sound amplitude and the squared value of sound frequency. In other words:

$$P \propto A^2, f^2$$

Therefore, in overall, we have:

$$I \propto A^2, f^2, \frac{1}{r^2}$$

Thus:

$$\frac{I_2}{I_1} = \left(\frac{A_2}{A_1}\right)^2 \times \left(\frac{f_2}{f_1}\right)^2 \times \left(\frac{r_1}{r_2}\right)^2$$

$$\Rightarrow \frac{I_2}{I_1} = (\quad)^2 \times (\quad)^2 \times (\quad)^2$$

$$\Rightarrow \frac{I_2}{I_1} = 9$$

Problem

13.19. At a point, the sound intensity is about 10^{-6} W/m^2. Calculate the amount of sound intensity level at this point. Herein, $I_0 = 10^{-12}$ W/m^2 which is the sound intensity reference.

Difficulty level ○ Easy ● Normal ○ Hard
Calculation amount ○ Small ● Normal ○ Large

1) -6 *db*
2) -60 *db*
3) 6 *db*
4) 60 *db*

Partially Solved Exercise

13.20. The power of a source of sound is about π *W*. Calculate the sound intensity level at 0.5 *m* distance. The amount of sound intensity reference is $I_0 = 10^{-12}$ W/m^2.

Solution

Based on the information given in the problem, we have:

$$P = \pi\ W/m^2$$

$$r = 0.5\ m$$

$$I_0 = 10^{-12}\ W/m^2$$

As we know, the sound intensity can be calculated as follows.

$$I = \frac{P}{4\pi r^2}$$

Herein, *I*, *P*, and *r* are the sound intensity (W/m^2), power of sound (*W*), and distance from the sound source (*m*), respectively.

$$I = \frac{(\quad)}{4\pi \times (\quad)^2} = (\quad)$$

Moreover, the sound intensity level can be calculated as follows.

$$\beta = 10 \log \frac{I}{I_0}$$

Herein, β, *I*, and I_0 are the sound intensity level (*db*), sound intensity (W/m^2), and sound intensity reference (W/m^2), respectively. Herein, $I_0 = 10^{-12}$ W/m^2.

Thus:

$$\beta = 10 \log \frac{(\quad)}{(\quad)}$$

$$\Rightarrow \beta = 10 \log(\quad)$$

$$\Rightarrow \beta = 120 \log(\quad)$$

$$\Rightarrow \beta = 120\ db$$

Notes

In this problem, the relation below has been used.

$$\frac{1}{a^{-b}} = a^b$$

$$a \log b^n = an \log b$$

$$\log 10 = 1$$

Problem

13.21. The amplitude of a sound source is tripled but its frequency and the distance from it are halved. Calculate the difference between the sound intensity levels ($\beta_2 - \beta_1$) concerned with the two tests.

Difficulty level ○ Easy ● Normal ○ Hard
Calculation amount ○ Small ● Normal ○ Large

1) 10 log 3 *db*
2) 20 log 9 *db*
3) 20 log 3 *db*
4) −10 log 9 *db*

Problem

13.22. Which one of the following choices is correct about the sound intensity level at a point if the distance from the sound source is tripled.

Difficulty level ○ Easy ● Normal ○ Hard
Calculation amount ○ Small ● Normal ○ Large

1) The sound intensity level rises about 9.54 *db*.
2) The sound intensity level drops about 9.54 *db*.
3) The sound intensity level rises about 4.77 *db*.
4) The sound intensity level drops about 4.77 *db*.

Partially Solved Exercise

13.23. Calculate the difference between the sound intensity levels ($\beta_2 - \beta_1$) if the amplitude, frequency and the distance from the sound source is halved.

Solution

Based on the information given in the problem, we have:

$$A_2 = 0.5A_1$$

$$f_2 = 0.5f_1$$

$$r_2 = 0.5r_1$$

The sound intensity level can be calculated as follows.

$$\beta = 10 \log \frac{I}{I_0}$$

Herein, β, I, and I_0 are the sound intensity level (db), sound intensity (W/m^2), and sound intensity reference (W/m^2), respectively. Herein, $I_0 = 10^{-12}\ W/m^2$.

In addition, the difference between two sound intensity levels, that is, I_2 and I_1 can be calculated as follows.

$$\Delta\beta = \beta_2 - \beta_1 = 10 \log \frac{I_2}{I_1}$$

Since $I \propto A^2, f^2, \frac{1}{r^2}$, we can conclude that:

$$\Delta\beta = 10 \log \left(\frac{A_2}{A_1} \times \frac{f_2}{f_1} \times \frac{r_1}{r_2}\right)^2$$

Hence:

$$\Delta\beta = 10 \log ((\quad) \times (\quad) \times (\quad))^2$$

$$\Rightarrow \Delta\beta = 10 \log (\quad)^2 = 20 \log(\quad)$$

$$\Rightarrow \Delta\beta = -20 \log 2\ db$$

Notes

In this problem, the relation below has been used.

$$a \log b^n = an \log b$$

$$\log\left(\frac{a}{b}\right) = \log a - \log b$$

$$\log 1 = 0$$

13.4 Doppler Effect

Problem

13.24. An observer at a velocity which is half of the speed of sound is approaching a stationary source of sound. Calculate the ratio of frequency that the person hears to the frequency of the source.

Difficulty level ○ Easy ● Normal ○ Hard

Calculation amount ○ Small ● Normal ○ Large

1) $\frac{1}{2}$
2) $\frac{2}{3}$
3) $\frac{3}{2}$
4) 2

Problem

13.25. An observer and a source of sound are moving toward each other that the velocity of each of them is half of sound speed. If the frequency of sound that the person hears is 180 Hz, calculate the frequency of the source.

Difficulty level ○ Easy ● Normal ○ Hard

Calculation amount ○ Small ● Normal ○ Large

1) 180 *Hz*
2) 540 *Hz*
3) 90 *Hz*
4) 60 *Hz*

Partially Solved Exercise

13.26. A source of sound, at half speed of sound, is approaching an observer that moves towards the source with the velocity of v_o. Calculate the speed of the observer if the frequency of sound that the observer hears is three times of the frequency of the source?

Solution

Based on the information given in the problem, we have:

$$v_s = 0.5v_{sound}$$

$$f_o = 3f_s$$

Based on Doppler effect, the frequency of sound that the observer hears (f_o in Hz) can be calculated as follows.

$$f_o = \frac{v_{sound} \pm v_o}{v_{sound} \pm v_s} f_s$$

Herein, v_{sound}, v_o, v_s, and f_s are the speed of sound (m/s), speed of observer (m/s), speed of sound source (m/s), and frequency of the sound source (Hz).

Moreover, the positive or negative sign for v_o and v_s is chosen based on the reference direction which is from the source to the observer.

Hence, by using Fig. 13.2, we can write:

$$(\qquad) = \frac{v_{sound} + (\qquad\qquad)}{v_{sound} + (\qquad\qquad)} f_s$$

$$\Rightarrow (\qquad) = \frac{v_{sound} + v_o}{(\qquad)v_{sound}}$$

$$\Rightarrow v_{sound} + v_o = (\qquad)v_{sound}$$

$$\Rightarrow v_o = 0.5v_{sound}$$

Fig. 13.2 The direction of movement of a source of sound and an observer to study Doppler effect

References

1. Rahmani-Andebili, M., General Physics I – Practice Problems, Methods, and Solutions, Springer Nature, 2025.
2. Rahmani-Andebili, M., Calculus III – Practice Problems, Methods, and Solutions, Springer Nature, 2023.
3. Rahmani-Andebili, M., Calculus II – Practice Problems, Methods, and Solutions, Springer Nature, 2023.
4. Rahmani-Andebili, M., Calculus I (2nd Ed.) – Practice Problems, Methods, and Solutions, Springer Nature, 2023.
5. Rahmani-Andebili, M., Precalculus (2nd Ed.) – Practice Problems, Methods, and Solutions, Springer Nature, 2024.

14 Transverse and Longitudinal Waves: Part B

Abstract
In this chapter, the problems of the thirteenth chapter are fully solved, in detail, step-by-step, and with different methods.

14.1 Transverse Waves

14.1. Two points are in-phase if the distance between them is an integer coefficient of wavelength [1–5]. In other words:

$$\Delta x = n\lambda, n\epsilon Z$$

Moreover, two points are called in-phase if the phase angle difference between them is an even coefficient of π. In other words:

$$\Delta\varphi = 2n\pi, n\epsilon Z$$

Therefore, as can be noticed from Fig. 14.1, Point E is in phase with Point A.

$$\Delta x = \lambda$$

$$\Delta\varphi = 2\pi$$

Choice (4) is the answer.

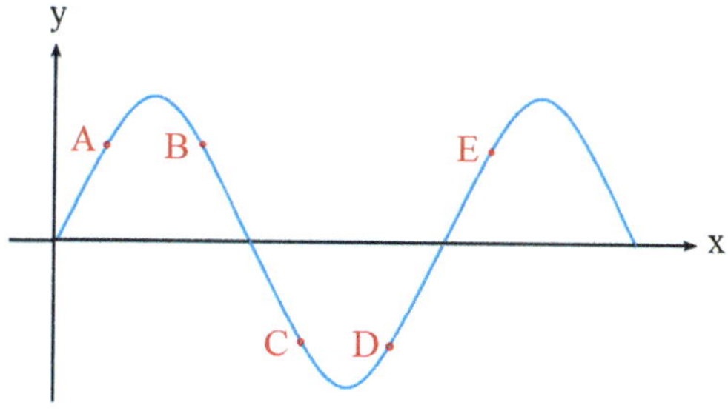

Fig. 14.1 A transverse wave

M. Rahmani-Andebili, *General Physics II*, https://doi.org/10.1007/978-3-031-92866-6_14

14.2. Based on the information given in the problem, we have:

$$T = 0.25\ s$$

$$\lambda = 10\ m$$

As we know, the relation below exists between the velocity (*m/s*), frequency (*Hz*), wavelength (*m*), and period (*s*) of a transverse wave.

$$v = f\lambda = \frac{\lambda}{T}$$

Therefore:

$$v = \frac{10}{0.25} \Rightarrow v = 40\ m/s$$

Choice (4) is the answer.

14.4. Based on the information given in the problem, we have:

$$f = 60\ Hz$$

$$\lambda = 0.5\ m$$

$$x = 200\ m$$

As we know, the relation below exists between the velocity (*m/s*), frequency (*Hz*), wavelength (*m*), and period (*s*) of a transverse wave.

$$v = f\lambda = \frac{\lambda}{T}$$

Thus:

$$v = 0.5 \times 60 = 30\ m/s$$

Moreover, as we know:

$$v = \frac{\Delta x}{\Delta t}$$

$$\Rightarrow \Delta t = \frac{\Delta x}{v} \Rightarrow \Delta t = \frac{200}{30} \Rightarrow \Delta t = 6.66\ s$$

Choice (1) is the answer.

14.6. Based on the information given in the problem, we have:

$$\frac{v_{water}}{v_{air}} = 4$$

As we know, the relation below exists between the velocity (m/s), frequency (Hz), wavelength (m), and period (s) of a transverse wave.

$$v = f\lambda = \frac{\lambda}{T}$$

As can be seen, the wavelength of a wave is proportional to its velocity. In other words:

$$\lambda \propto v$$

Thus:

$$\frac{\lambda_{water}}{\lambda_{air}} = \frac{v_{water}}{v_{air}}$$

$$\Rightarrow \frac{\lambda_{water}}{\lambda_{air}} = 4$$

Choice (3) is the answer.

14.8. Based on the information given in the problem, we have:

$$u(t) = 0.1 \sin\left(\pi t - \frac{\pi}{3}\right) \tag{14.1}$$

$$\Delta t = 0.02\ s \tag{14.2}$$

From (14.1), the angular frequency of the wave can be extracted as follows.

$$\omega = \pi$$

In a wave equation, the relation between the phase angle change ($\Delta\varphi$), angular frequency (ω), and time change (Δt) is as follows.

$$\Delta\varphi = \omega \Delta t$$

Therefore:

$$\Delta\varphi = \pi \times 0.05 \Rightarrow \Delta\varphi = 0.05\pi\ rad$$

A phase angle can be converted from radian to degree as follows.

$$\varphi_{deg} = \varphi_{rad} \times \frac{180}{\pi}$$

Thus:

$$\Delta\varphi = 0.05\pi \times \frac{180}{\pi}$$

$$\Rightarrow \Delta\varphi = 9^\circ$$

Choice (1) is the answer.

Notes

In this problem, the relation below has been used.

The general form of a wave equation is as follows.

$$u(t) = A\sin(\omega t + \varphi_0)$$

Herein, A, ω, and φ_0 are the amplitude, angular frequency, and primary phase angle of the wave, respectively.

14.2 Longitudinal Waves

14.9. Based on the information given in the problem, we have:

$$l = 2\ m$$

$$m = 0.2\ kg$$

$$F = 10\ N$$

The propagation velocity of transverse waves in a string tightened between two points can be calculated by one of the following relations.

$$v = \sqrt{\frac{Fl}{m}}$$

$$v = \sqrt{\frac{F}{\rho A}}$$

Herein, F, l, m, ρ, A and are the tension force (N), length (m), mass (kg), density (kg/m^3), and surface area of the string (m^2), respectively.

Thus:

$$v = \sqrt{\frac{10 \times 2}{0.2}} \Rightarrow v = 10\ m/s$$

Choice (4) is the answer.

14.10. The propagation velocity of the waves will remain constant because it does not depend on the physical conditions of the source. Hence, the ratio of propagation velocities of transverse waves will be one. Choice (2) is the answer.

14.11. Based on the information given in the problem, we have:

$$\rho = 2\ gr/cm^3$$

$$A = 2\ mm^2$$

$$F = 100\ N$$

The propagation velocity of transverse waves in a string tightened between two points can be calculated by one of the following relations.

$$v = \sqrt{\frac{Fl}{m}}$$

$$v = \sqrt{\frac{F}{\rho A}}$$

Herein, F, l, m, ρ, A and are the tension force (N), length (m), mass (kg), density (kg/m^3), and surface area of the string (m^2), respectively.

Therefore:

$$v = \sqrt{\frac{100}{2 \times 10^3 \times 2 \times 10^{-6}}}$$

$$\Rightarrow v = \sqrt{\frac{10^5}{4}} \Rightarrow v = 158.1\ m/s$$

Choice (4) is the answer.

Notes

In this problem, the relation below has been used.

$$1\ \frac{gr}{cm^3} = 10^3\ \frac{kg}{m^3}$$

14.12. Based on the information given in the problem, we have:

$$F_1 = F_2$$

$$A_1 = 2A_2$$

The propagation velocity of transverse waves in a string tightened between two points can be calculated by one of the following relations.

$$v=\sqrt{\frac{Fl}{m}}$$

$$v=\sqrt{\frac{F}{\rho A}}$$

Herein, F, l, m, ρ, A and are the tension force (N), length (m), mass (kg), density (kg/m^3), and surface area of the string (m^2), respectively.

Hence:

$$\frac{v_1}{v_2}=\sqrt{\frac{F_1}{F_2}\times\frac{\rho_2}{\rho_1}\times\frac{A_2}{A_1}}$$

$$\Rightarrow\frac{v_1}{v_2}=\sqrt{1\times1\times\frac{1}{2}}$$

$$\Rightarrow\frac{v_1}{v_2}=\frac{1}{\sqrt{2}}=\frac{1}{\sqrt{2}}\times\frac{\sqrt{2}}{\sqrt{2}}$$

$$\Rightarrow\frac{v_1}{v_2}=\frac{\sqrt{2}}{2}$$

Choice (2) is the answer.

14.3 Sound Intensity and Sound Intensity Level

14.14. Frequency is one of the inherent features of a sound source. Thus, the frequency of sound in water and air is the same Choice (1) is the answer.

14.15. The speed of sound depends on the medium. The speed of sound in water is larger than the one in air because water is denser than air. Choice (2) is the answer.

14.16. Based on the information given in the problem, we have:

$$P=100\ W$$

$$I_{pain}=1\ W/m^2$$

As we know, the sound intensity can be calculated as follows.

$$I = \frac{P}{4\pi r^2}$$

Herein, I, P, and r are the sound intensity (W/m^2), power of sound (W), and distance from the sound source (m), respectively.

Hence:

$$1 = \frac{100}{4\pi r^2} \Rightarrow r^2 = \frac{100}{4\pi}$$

$$\Rightarrow r^2 = 7.96 \Rightarrow r = 2.82\ m$$

Choice (2) is the answer.

14.17. Based on the information given in the problem, we have:

$$r_2 = 3r_1$$

As we know, the sound intensity can be calculated as follows.

$$I = \frac{P}{4\pi r^2}$$

Herein, I, P, and r are the sound intensity (W/m^2), power of sound (W), and distance from the sound source (m), respectively.

Moreover, the sound power is proportional to the squared value of sound amplitude and the squared value of sound frequency. In other words:

$$P \propto A^2, f^2$$

Therefore, in overall, we have:

$$I \propto A^2, f^2, \frac{1}{r^2}$$

Hence:

$$\frac{I_2}{I_1} = \left(\frac{r_1}{r_2}\right)^2$$

$$\Rightarrow \frac{I_2}{I_1} = \left(\frac{1}{3}\right)^2 \Rightarrow \frac{I_2}{I_1} = \frac{1}{9}$$

Choice (3) is the answer.

14.19. Based on the information given in the problem, we have:

$$I = 10^{-6}\ W/m^2$$

$$I_0 = 10^{-12}\ W/m^2$$

The sound intensity level is calculated as follows.

$$\beta = 10 \log \frac{I}{I_0}$$

Herein, β, I, and I_0 are the sound intensity level (*db*), sound intensity (*W*/*m*^{2}), and sound intensity reference (*W*/*m*^{2}), respectively. Herein, $I_0 = 10^{-12}$ *W*/*m*^{2}.

Thus:

$$\beta = 10 \log \frac{10^{-6}}{10^{-12}}$$

$$\Rightarrow \beta = 10 \log 10^6$$

$$\Rightarrow \beta = 60 \log 10$$

$$\Rightarrow \beta = 60\ db$$

Choice (4) is the answer.

Notes

In this problem, the relation below has been used.

$$\frac{a^m}{a^n} = a^{m-n}$$

$$a \log b^n = an \log b$$

$$\log 10 = 1$$

14.21. Based on the information given in the problem, we have:

$$A_2 = 3A_1$$

$$f_2 = \frac{1}{2} f_1$$

$$r_2 = \frac{1}{2} r_1$$

The sound intensity level can be calculated as follows.

$$\beta = 10 \log \frac{I}{I_0}$$

Herein, β, I, and I_0 are the sound intensity level (db), sound intensity (W/m^2), and sound intensity reference (W/m^2), respectively. Herein, $I_0 = 10^{-12}\ W/m^2$.

In addition, the difference between two sound intensity levels, that is, I_2 and I_1 can be calculated as follows.

$$\Delta\beta = \beta_2 - \beta_1 = 10 \log \frac{I_2}{I_1}$$

Since $I \propto A^2, f^2, \frac{1}{r^2}$, we can conclude that:

$$\Delta\beta = 10 \log \left(\frac{A_2}{A_1} \times \frac{f_2}{f_1} \times \frac{r_1}{r_2}\right)^2$$

Hence:

$$\Delta\beta = 10 \log \left(3 \times \frac{1}{2} \times 2\right)^2$$

$$\Rightarrow \Delta\beta = 10 \log 3^2$$

$$\Rightarrow \Delta\beta = 20 \log 3\ db$$

Choice (3) is the answer.

Notes

In this problem, the relation below has been used.

$$a \log b^n = an \log b$$

14.22. Based on the information given in the problem, we have:

$$r_2 = 3r_1$$

The sound intensity level can be calculated as follows.

$$\beta = 10 \log \frac{I}{I_0}$$

Herein, β, I, and I_0 are the sound intensity level (db), sound intensity (W/m^2), and sound intensity reference (W/m^2), respectively. Herein, $I_0 = 10^{-12}\ W/m^2$.

In addition, the difference between two sound intensity levels, that is, I_2 and I_1 can be calculated as follows.

$$\Delta\beta = \beta_2 - \beta_1 = 10 \log \frac{I_2}{I_1}$$

Since $I \propto A^2, f^2, \frac{1}{r^2}$, we can conclude that:

$$\Delta\beta = 10 \log \left(\frac{A_2}{A_1} \times \frac{f_2}{f_1} \times \frac{r_1}{r_2} \right)^2$$

Hence:

$$\Delta\beta = 10 \log \left(\frac{1}{3} \right)^2$$

$$\Rightarrow \Delta\beta = 20 \log \left(\frac{1}{3} \right) \Rightarrow \Delta\beta = 20(\log 1 - \log 3)$$

$$\Rightarrow \Delta\beta = 20\ (0 - \log 3) \Rightarrow \Delta\beta = -20 \log 3$$

$$\Rightarrow \Delta\beta = -9.54\ db$$

Thus, the sound intensity level drops about 9.54 *db*. Choice (2) is the answer.

Notes

In this problem, the relation below has been used.

$$a \log b^n = an \log b$$

$$\log \left(\frac{a}{b} \right) = \log a - \log b$$

$$\log 1 = 0$$

$$\log 3 = 0.477$$

14.4 Doppler Effect

14.24. Based on the information given in the problem, we have:

$$v_o = 0.5 v_{sound}$$

$$v_s = 0$$

Based on Doppler effect, the frequency of sound that the observer hears (f_o in Hz) can be calculated as follows.

$$f_o = \frac{v_{sound} \pm v_o}{v_{sound} \pm v_s} f_s$$

Herein, v_{sound}, v_o, v_s, and f_s are the speed of sound (m/s), speed of observer (m/s), speed of sound source (m/s), and frequency of the sound source (Hz).

Moreover, the positive or negative sign for v_o and v_s is chosen based on the reference direction which is from the source to the observer.

Hence, by using Fig. 14.2, we can write:

$$f_o = \frac{v_{sound} + 0.5v_{sound}}{v_{sound}} f_s$$

$$\Rightarrow \frac{f_o}{f_s} = \frac{1.5v_{sound}}{v_{sound}} \Rightarrow \frac{f_o}{f_s} = 1.5$$

Choice (3) is the answer.

Fig. 14.2 The direction of movement of a source of sound and an observer to study Doppler effect

---

14.25. Based on the information given in the problem, we have:

$$v_o = 0.5v_{sound}$$

$$v_s = 0.5v_{sound}$$

$$f_o = 180\ Hz$$

Based on Doppler effect, the frequency of sound that the observer hears (f_o in Hz) can be calculated as follows.

$$f_o = \frac{v_{sound} \pm v_o}{v_{sound} \pm v_s} f_s$$

Herein, v_{sound}, v_o, v_s, and f_s are the speed of sound (m/s), speed of observer (m/s), speed of sound source (m/s), and frequency of the sound source (Hz).

Moreover, the positive or negative sign for v_o and v_s is chosen based on the reference direction which is from the source to the observer.
Hence, by using Fig. 14.3, we can write:

$$180 = \frac{v_{sound} + 0.5v_{sound}}{v_{sound} - 0.5v_{sound}} f_s$$

$$\Rightarrow f_s = \frac{0.5v_{sound}}{1.5v_{sound}} \times 180$$

$$\Rightarrow f_s = 60\ Hz$$

Choice (4) is the answer.

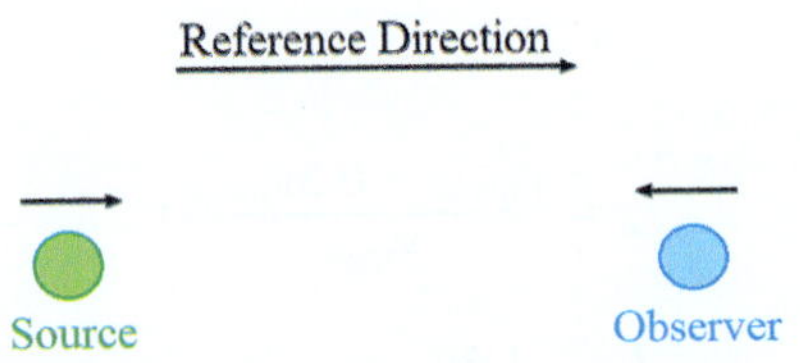

Fig. 14.3 The direction of movement of a source of sound and an observer to study Doppler effect

References

1. Rahmani-Andebili, M., General Physics I – Practice Problems, Methods, and Solutions, Springer Nature, 2025.
2. Rahmani-Andebili, M., Calculus III – Practice Problems, Methods, and Solutions, Springer Nature, 2023.
3. Rahmani-Andebili, M., Calculus II – Practice Problems, Methods, and Solutions, Springer Nature, 2023.
4. Rahmani-Andebili, M., Calculus I (2nd Ed.) – Practice Problems, Methods, and Solutions, Springer Nature, 2023.
5. Rahmani-Andebili, M., Precalculus (2nd Ed.) – Practice Problems, Methods, and Solutions, Springer Nature, 2024.

15 Light, Mirrors, and Lenses: Part A

Abstract

In this chapter, the basic and advanced problems of Light, Mirrors, and Lenses are studied. The subjects include Reflection, Refraction, Flat Mirror, Concave Mirror, Convex Mirror, Convex Lens, and Concave Lens. Herein, different types of problems and exercises are presented that are categorized as follows.

- ***Problems with detailed solution***: They have been designed to teach students the subjects in detail. Moreover, they have been categorized in different levels based on their difficulty levels (easy, normal, and hard) and calculation amounts (small, normal, and large).
- ***Partially solved exercises***: They have been designed to encourage students to practice problems while guiding them through the problem-solving procedure and hinting the required formulas.
- ***Exercises with final answer***: They have been designed to encourage students to practice more by themselves while hinting them by the final answer as well as to help instructors to give tests or quizzes.

15.1 Reflection and Refraction

Problem

15.1. A ray radiates from a prism at the angle of 45° and then enter to water. Which one of the following choices is correct about the refracted ray. The refractive index of prism and water are 1.5 and 1.33, respectively [1–5].

Difficulty level ○ Easy ● Normal ○ Hard

Calculation amount ● Small ○ Normal ○ Large

1) The refracted ray enters the water and moves away from the normal axis.
2) The refracted ray tangentially leaves the prism.
3) The refracted ray enters the water and comes close to the normal axis.
4) The incident ray is reflected from the water.

Partially Solved Exercise

15.2. In Problem 15.1, calculate the angle of refraction.

Solution

Based on the information given in the problem, we have:

$$\theta_1 = 45°$$

M. Rahmani-Andebili, *General Physics II*, https://doi.org/10.1007/978-3-031-92866-6_15

$$n_1 = 1.5$$

$$n_2 = 1.33$$

As we know, the relation between the refractive index of two environments, sine value of incident angle, and sine value of refracted angle is as follows (Fig. 15.1).

$$\frac{\sin\theta_1}{\sin\theta_2} = \frac{n_2}{n_1}$$

Hence:

$$\frac{\sin(\quad)^\circ}{\sin\theta_2} = \frac{(\quad)}{(\quad)}$$

$$\Rightarrow \sin\theta_2 = \frac{(\quad)}{(\quad)} \times (\quad)$$

$$\Rightarrow \sin\theta_2 = (\quad)$$

$$\Rightarrow \theta_2 = 53.1^\circ$$

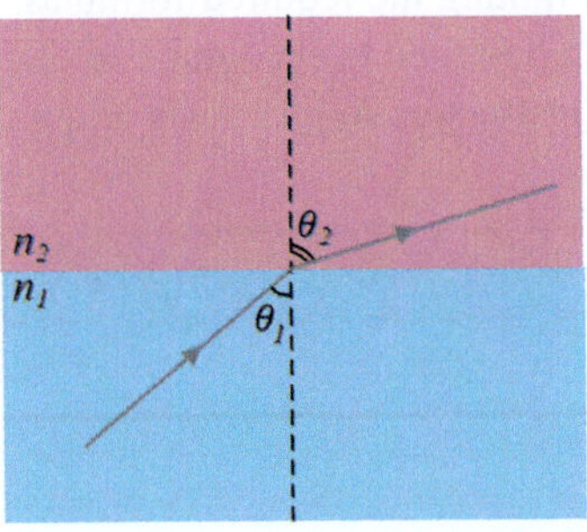

Fig. 15.1 The incident and refracted rays at the boundary of two media

Notes

In this problem, the relation below has been used.

$$\sin 45^\circ = \frac{\sqrt{2}}{2}$$

Problem

15.3. A ray radiates from water of a pool at the angle of 48.6° and then enter to air. Calculate the angle of refraction. The refractive index of water is 1.33.

Difficulty level ○ Easy ● Normal ○ Hard
Calculation amount ● Small ○ Normal ○ Large

1) 48.6°
2) 45°
3) 50°
4) 90°

Problem

15.4. A ray radiates from water of a pool at the angle of 50° and then enter to air. Calculate the angle of refraction. The refractive index of water is 1.33.

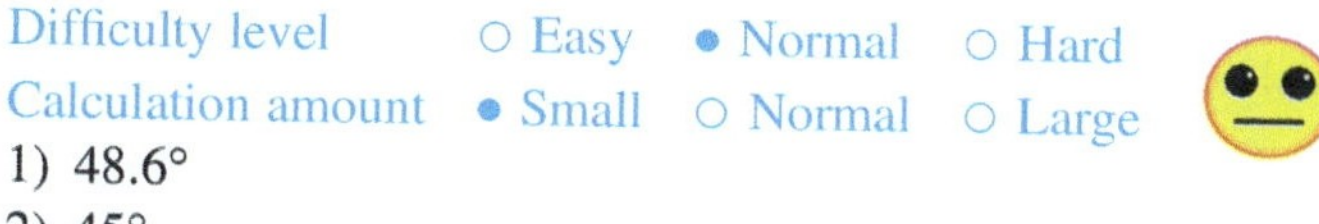

1) 48.6°
2) 45°
3) 50°
4) 90°

Problem

15.5. Consider a pool with 2 *m* depth which is full of water. Determine the apparent depth of the pool if the refractive index of water is 1.33.

Difficulty level ○ Easy ● Normal ○ Hard
Calculation amount ● Small ○ Normal ○ Large

1) 2 *m*
2) 2.66 *m*
3) 1.33 *m*
4) 1.5 *m*

Problem

15.6. Consider a diver who dives in the sea. The diver looks at a ship with 7.52 *m* height. Determine the apparent height of the ship that the diver see if the refractive index of seawater is 1.33.

Difficulty level ○ Easy ● Normal ○ Hard
Calculation amount ● Small ○ Normal ○ Large

1) 7.51 *m*
2) 5.64 *m*
3) 10 *m*
4) 11.33 *m*

15.2 Flat Mirror

Problem

15.7. The sum of the angles of incident and reflection of a ray from a flat mirror is 150°. Calculate the angle between the reflected ray and the mirror.

Difficulty level ● Easy ○ Normal ○ Hard

Calculation amount ● Small ○ Normal ○ Large

1) 75°
2) 15°
3) 10°
4) 20°

Problem

15.8. Consider the two flat mirrors shown in Fig. 15.2 and calculate the angle θ.

Difficulty level ○ Easy ● Normal ○ Hard

Calculation amount ○ Small ● Normal ○ Large

1) 50°
2) 40°
3) 20°
4) 10°

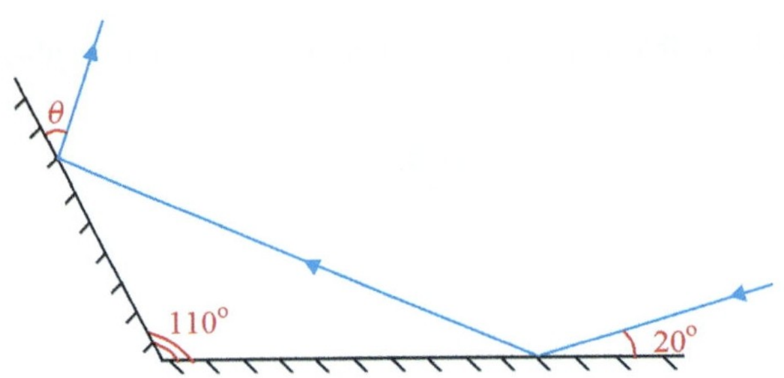

Fig. 15.2 The incident and reflected rays related to two flat mirrors

Problem

15.9. If an object is moved away from a flat mirror about one meter, how far will it move away from its image?

Difficulty level ○ Easy ● Normal ○ Hard

Calculation amount ● Small ○ Normal ○ Large

1) 1 m
2) 0.5 m
3) 2 m
4) 4 m

Problem

15.10. If an object is placed between two flat mirrors that the angle between them is α. Which one of the following choices is correct about the number of images in these two mirrors.

Difficulty level ○ Easy ● Normal ○ Hard
Calculation amount ● Small ○ Normal ○ Large

1) The number of images will increase if α increases.
2) The number of images will decrease if α increases.
3) The number of images is two for any value of α.
4) The number of images is infinite for any value of α.

Exercise

15.11. An object is placed between two flat mirrors that the angle between them is 30°. Calculate the number of images that are created in these two mirrors.

Final Answer

$n = 11$

15.3 Concave Mirror

Problem

15.12. Which one of the choices illustrated in Fig. 15.3 shows the correct ray reflected from the concave mirror?

Difficulty level ● Easy ○ Normal ○ Hard
Calculation amount ● Small ○ Normal ○ Large

1) Reflected ray (1)
2) Reflected ray (2)
3) Reflected ray (3)
4) Reflected ray (4)

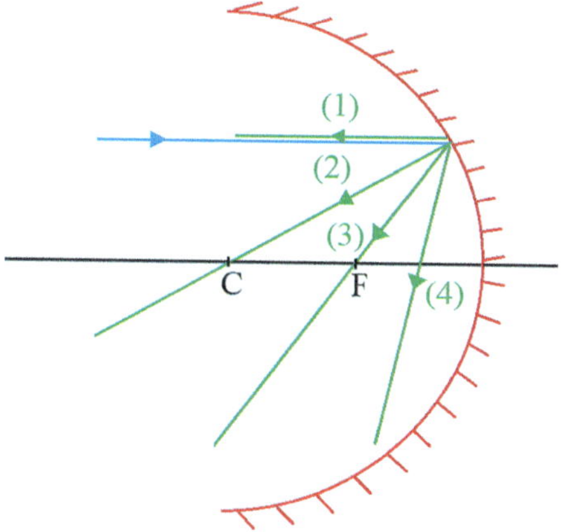

Fig. 15.3 The incident and reflected rays related to a concave mirror

Problem

15.13. Which one of the choices illustrated in Fig. 15.4 shows the correct ray reflected from the concave mirror?

Difficulty level ● Easy ○ Normal ○ Hard

Calculation amount ● Small ○ Normal ○ Large

1) Reflected ray (1)
2) Reflected ray (2)
3) Reflected ray (3)
4) Reflected ray (4)

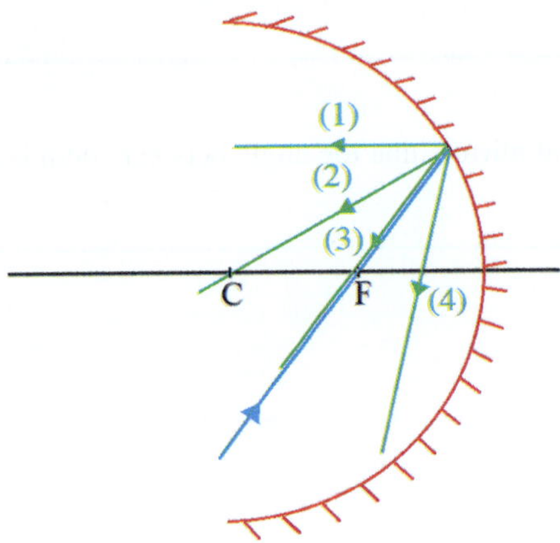

Fig. 15.4 The incident and reflected rays related to a concave mirror

Problem

15.14. Which one of the choices illustrated in Fig. 15.5 shows the correct ray reflected from the concave mirror?

Difficulty level ● Easy ○ Normal ○ Hard

Calculation amount ● Small ○ Normal ○ Large

1) Reflected ray (1)
2) Reflected ray (2)
3) Reflected ray (3)
4) Reflected ray (4)

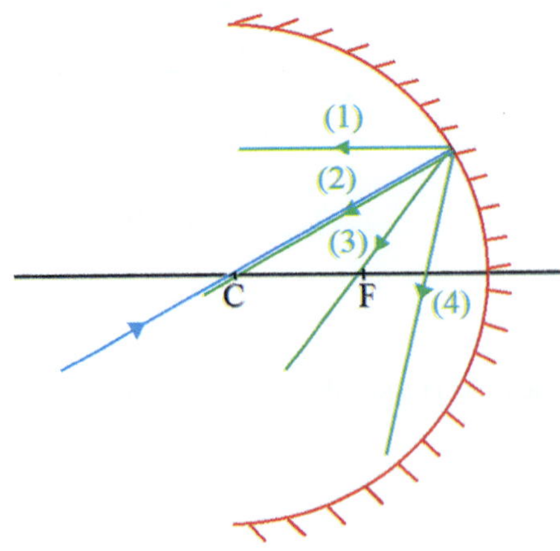

Fig. 15.5 The incident and reflected rays related to a concave mirror

Problem

15.15. In the system shown in Fig. 15.6, calculate the distance between the points A and B. Herein, f_1 and f_2 are the focal lengths of the first and second concave mirrors.

Difficulty level ○ Easy ○ Normal ● Hard
Calculation amount ● Small ○ Normal ○ Large

1) $f_1 + f_2$
2) $f_1 - f_2$
3) $2f_1 + f_2$
4) $f_1 + 2f_2$

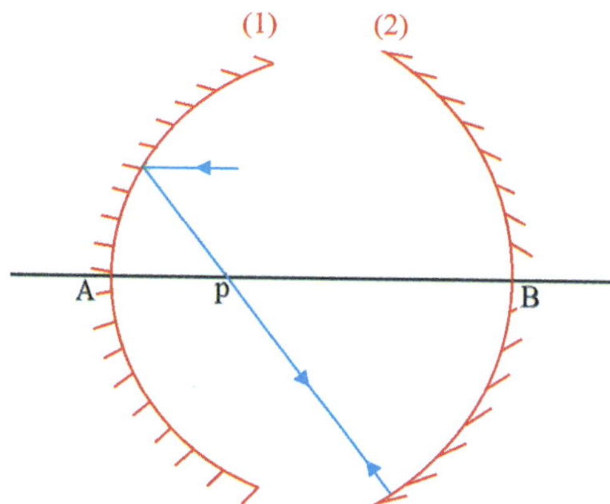

Fig. 15.6 The incident and reflected rays concerned with two parallel concave mirrors

Problem

15.16. Which one of the following choices is wrong about the image of a concave mirror if the object is placed in the focal distance.

Difficulty level ○ Easy ● Normal ○ Hard
Calculation amount ● Small ○ Normal ○ Large

1) The image is smaller than the object.
2) The image is virtual.
3) The image is straight.
4) The image is behind the concave mirror.

Problem

15.17. Which one of the following choices is wrong about the image of a concave mirror if the object is placed on focal point (focus).

Difficulty level ○ Easy ● Normal ○ Hard
Calculation amount ● Small ○ Normal ○ Large

1) The image is larger than the object.
2) The image is real.
3) The image is inverted.
4) The image is on the focal point (focus).

Problem

15.18. Which one of the following choices is wrong about the image of a concave mirror if the object is placed between the focal point (focus) and the center of the mirror.

Difficulty level ○ Easy ● Normal ○ Hard

Calculation amount ● Small ○ Normal ○ Large

1) The image is larger than the object.
2) The image is virtual.
3) The image is inverted.
4) The image is on the object side but beyond the center of the concave mirror.

Problem

15.19. Which one of the following choices is wrong about the image of a concave mirror if the object is on the center of the mirror.

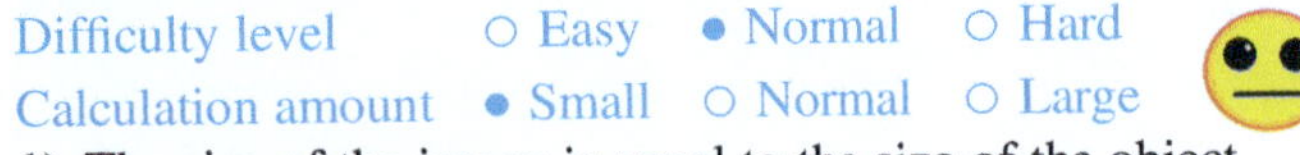

1) The size of the image is equal to the size of the object.
2) The image is real.
3) The image is straight.
4) The image is on the center of the concave mirror.

Problem

15.20. Which one of the following choices is wrong about the image of a concave mirror if the object is placed beyond the center of the mirror.

1) The image is smaller than the object.
2) The image is real.
3) The image is inverted.
4) The image is on the object side but beyond the center of the concave mirror.

Problem

15.21. Which one of the following choices is wrong about the image of a concave mirror if the object is at infinity.

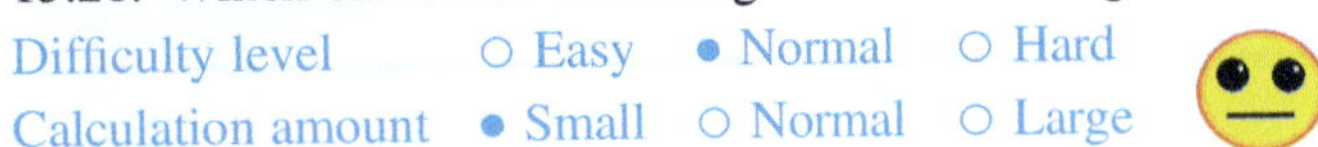

1) The image is larger than the object.
2) The image is real.
3) The image is inverted.
4) The image is on the focal point (focus).

Problem

15.22. If an object is moved from the center of a concave mirror to its focal point (focus), along the principal axis of the mirror, how will the image move?

Difficulty level ○ Easy ● Normal ○ Hard

Calculation amount ● Small ○ Normal ○ Large

1) The image will move from the focus to the center of the mirror.
2) The image will move from the center of the mirror to infinity.
3) The image will move from the center to the focus of the mirror.
4) The image will move from infinity to the center of the mirror.

Problem

15.23. The radius a concave mirror is 40 *cm*. Calculate the magnification of the mirror when the object is placed at 30 *cm* away from the mirror on its principal axis.

Difficulty level ○ Easy ● Normal ○ Hard

Calculation amount ○ Small ● Normal ○ Large

1) 2
2) $\frac{1}{2}$
3) $\frac{4}{3}$
4) $\frac{3}{4}$

Partially Solved Exercise

15.24. The radius a concave mirror is 50 *cm*. Calculate the distance of the image from the mirror when the object is placed at 10 *cm* away from the mirror on its principal axis.

Solution

Based on the information given in the problem, we have:

$$R = 50\ cm \Rightarrow f = 25\ cm$$

$$p = 10\ cm$$

As we know, for a concave mirror, the relation between the focus (f), distance of an object from the mirror (p), and distance of the image from the mirror (q) is as follows.

$$\frac{1}{p} + \frac{1}{q} = \frac{1}{f}$$

Herein, f, p, and q are positive quantities.

Therefore:

$$\frac{1}{(\quad)} + \frac{1}{q} = \frac{1}{(\quad)}$$

$$\Rightarrow \frac{1}{q} = \frac{1}{(\quad)} - \frac{1}{(\quad)} = \frac{(\quad)-(\quad)}{(\quad)} = (\quad)$$

$$\Rightarrow q = -16.6\ cm$$

The negative value of q means that the image is virtual and is created on the back of the mirror.

15.4 Convex Mirror

Problem

15.25. Which one of the choices illustrated in Fig. 15.7 shows the correct ray reflected from the convex mirror?

Difficulty level ● Easy ○ Normal ○ Hard

Calculation amount ● Small ○ Normal ○ Large

1) Reflected ray (1)
2) Reflected ray (2)
3) Reflected ray (3)
4) Reflected ray (4)

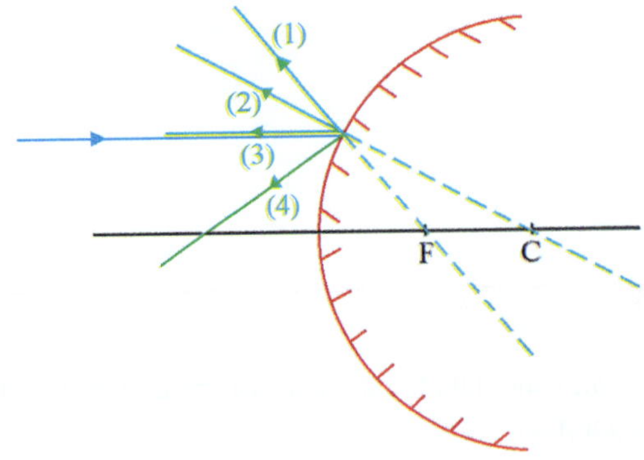

Fig. 15.7 The incident and reflected rays related to a convex mirror

Problem

15.26. Which one of the choices illustrated in Fig. 15.8 shows the correct ray reflected from the convex mirror?

Difficulty level ● Easy ○ Normal ○ Hard

Calculation amount ● Small ○ Normal ○ Large

1) Reflected ray (1)
2) Reflected ray (2)
3) Reflected ray (3)
4) Reflected ray (4)

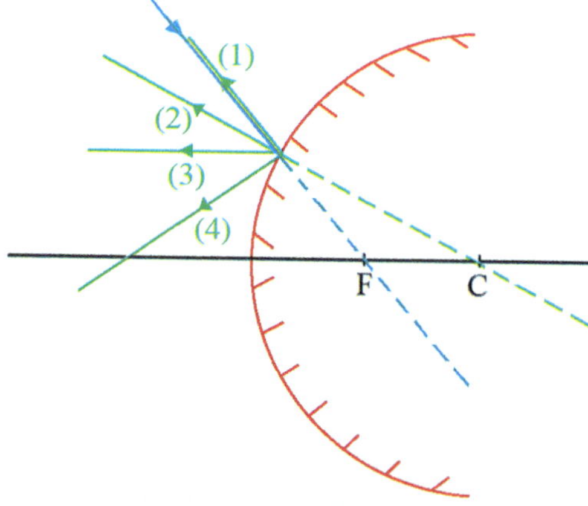

Fig. 15.8 The incident and reflected rays related to a convex mirror

Problem

15.27. Which one of the choices illustrated in Fig. 15.9 shows the correct ray reflected from the convex mirror?

Difficulty level ● Easy ○ Normal ○ Hard

Calculation amount ● Small ○ Normal ○ Large

1) Reflected ray (1)
2) Reflected ray (2)
3) Reflected ray (3)
4) Reflected ray (4)

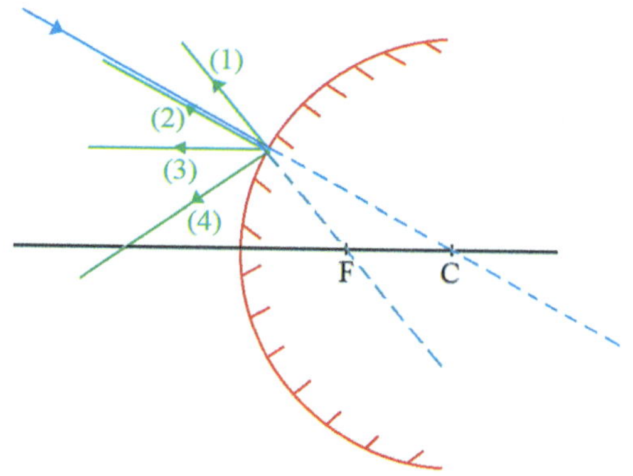

Fig. 15.9 The incident and reflected rays related to a convex mirror

Problem

15.28. Which one of the following choices is wrong about the image of a convex mirror.

Difficulty level ○ Easy ● Normal ○ Hard

Calculation amount ● Small ○ Normal ○ Large

1) The image is always smaller than the object.
2) The image is always virtual.
3) The image is always straight.
4) The image is always on the object side and on the focal distance.

Problem

15.29. The size of the image of an object on a convex mirror is half the size of the image. Calculate the distance of the object from the mirror if the radius of the mirror is 28 *cm*.

Difficulty level ○ Easy ○ Normal ● Hard
Calculation amount ○ Small ● Normal ○ Large

1) 7 *cm*
2) 14 *cm*
3) 28 *cm*
4) 56 *cm*

Problem

15.30. The focal distance of a convex mirror is 20 *cm*. Calculate the distance of the object from the mirror for the magnification of 0.4.

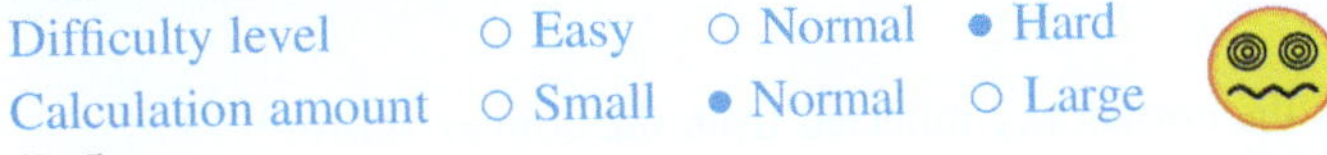
Difficulty level ○ Easy ○ Normal ● Hard
Calculation amount ○ Small ● Normal ○ Large

1) 5 *cm*
2) 10 *cm*
3) 20 *cm*
4) 30 *cm*

15.5 Convex Lens

Problem

15.31. Which one of the choices illustrated in Fig. 15.10 shows the correct ray refracted in the convex lens?

Difficulty level ● Easy ○ Normal ○ Hard
Calculation amount ● Small ○ Normal ○ Large

1) Reflected ray (1)
2) Reflected ray (2)
3) Reflected ray (3)
4) Reflected ray (4)

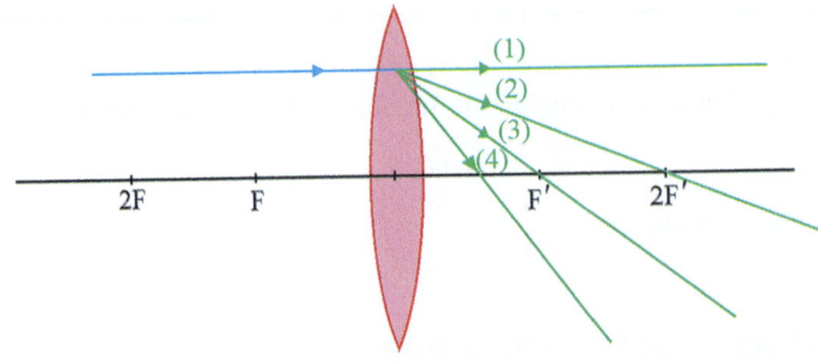

Fig. 15.10 The incident and refracted rays related to a convex lens

Problem

15.32. Which one of the choices illustrated in Fig. 15.11 shows the correct ray refracted in the convex lens?

Difficulty level ● Easy ○ Normal ○ Hard

Calculation amount ● Small ○ Normal ○ Large

1) Reflected ray (1)
2) Reflected ray (2)
3) Reflected ray (3)
4) Reflected ray (4)

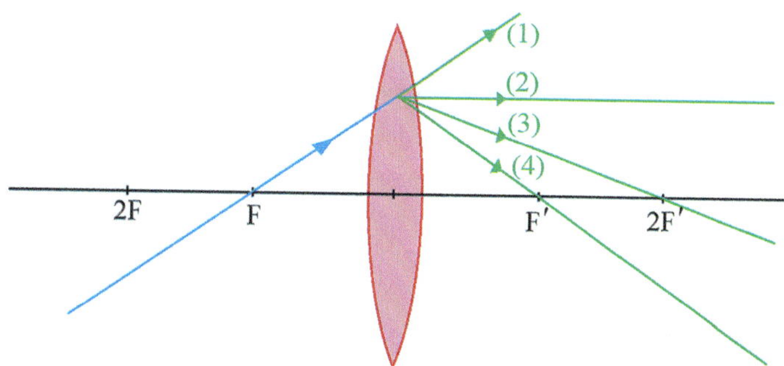

Fig. 15.11 The incident and refracted rays related to a convex lens

Problem

15.33. Which one of the choices illustrated in Fig. 15.12 shows the correct ray refracted in the convex lens?

Difficulty level ● Easy ○ Normal ○ Hard

Calculation amount ● Small ○ Normal ○ Large

1) Reflected ray (1)
2) Reflected ray (2)
3) Reflected ray (3)
4) Reflected ray (4)

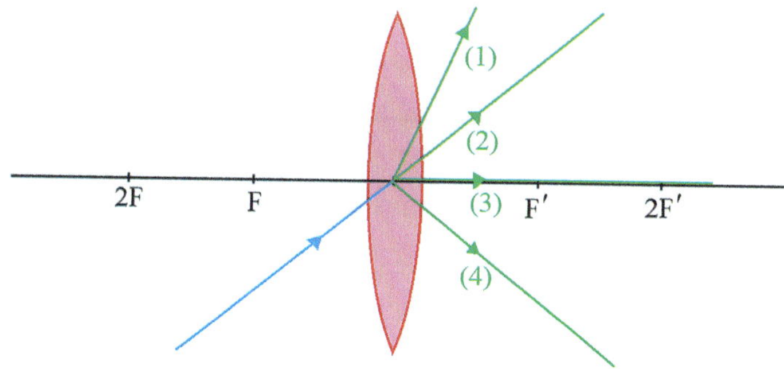

Fig. 15.12 The incident and refracted rays related to a convex lens

Problem

15.34. Which one of the following choices is wrong about the image of a convex lens if the object is placed in the focal distance.

Difficulty level ○ Easy ● Normal ○ Hard
Calculation amount ● Small ○ Normal ○ Large

1) The image is larger than the object.
2) The image is real.
3) The image is straight.
4) The image is on the object side.

Problem

15.35. Which one of the following choices is wrong about the image of a convex lens if the object is placed on focal point (focus).

1) The image is larger than the object.
2) The image is virtual.
3) The image is inverted.
4) The image is at infinity.

Problem

15.36. Which one of the following choices is wrong about the image of a convex lens if the object is placed between the focal point (focus) and the center of the curvature of the lens.

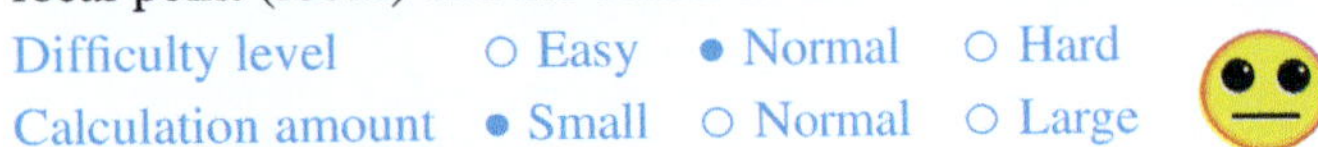

1) The image is larger than the object.
2) The image is real.
3) The image is straight.
4) The image is on the other side of the lens and beyond the center of the curvature of the lens.

Problem

15.37. Which one of the following choices is wrong about the image of a convex lens if the object is placed on the center of the curvature of the lens.

1) The size of the image is equal to the size of the object.
2) The image is real.
3) The image is inverted.
4) The image is on the same side as the object is, and is created on the center of the curvature of the lens.

Problem

15.38. Which one of the following choices is wrong about the image of a convex lens if the object is placed beyond the center of the curvature of the lens.

Difficulty level ○ Easy ● Normal ○ Hard
Calculation amount ● Small ○ Normal ○ Large

1) The image is larger than the object.
2) The image is real.
3) The image is inverted.
4) The image is on the other side of the lens and between the focal point and the center of the curvature of the lens.

Problem

15.39. Which one of the following choices is wrong about the image of a convex lens if the object is at infinity.

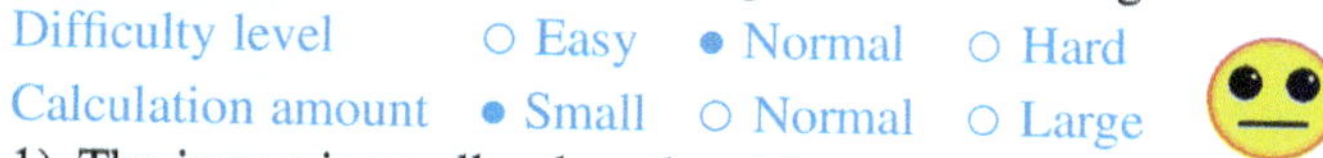

1) The image is smaller than the object.
2) The image is real.
3) The image is inverted.
4) The image is at infinity.

Problem

15.40. The focal point (focus) a convex lens is 20 *cm*. Calculate the magnification of the lens when the object is placed at 30 *cm* away from the lens on its principal axis.

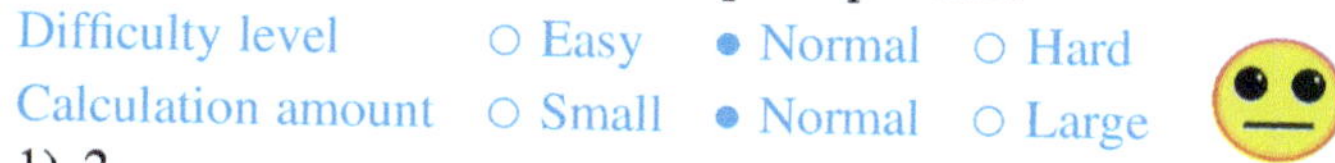

1) 2
2) $\frac{1}{2}$
3) $\frac{4}{3}$
4) $\frac{3}{4}$

Partially Solved Exercise

15.41. The focal point (focus) a convex lens 25 *cm*. Calculate the distance of the image from the lens when the object is placed at 30 *cm* away from the lens on its principal axis.

Solution

Based on the information given in the problem, we have:

$$f = 25\ cm$$

$$p = 30\ cm$$

As we know, for a convex lens, the relation between the focus (f), distance of an object from the lens (p), and distance of the image from the lens (q) is as follows.

$$\frac{1}{p} + \frac{1}{q} = \frac{1}{f}$$

Herein, f, p, and q are positive quantities.

Therefore:

$$\frac{1}{(\quad)} + \frac{1}{q} = \frac{1}{(\quad)}$$

$$\Rightarrow \frac{1}{q} = \frac{1}{(\quad)} - \frac{1}{(\quad)} = \frac{(\quad) - (\quad)}{(\quad)} = (\quad)$$

$$\Rightarrow q = 150\ cm$$

15.6 Concave Lens

Problem

15.42. Which one of the choices illustrated in Fig. 15.13 shows the correct ray refracted in the concave lens?

Difficulty level • Easy ○ Normal ○ Hard 

Calculation amount • Small ○ Normal ○ Large 

1) Reflected ray (1)
2) Reflected ray (2)
3) Reflected ray (3)
4) Reflected ray (4)

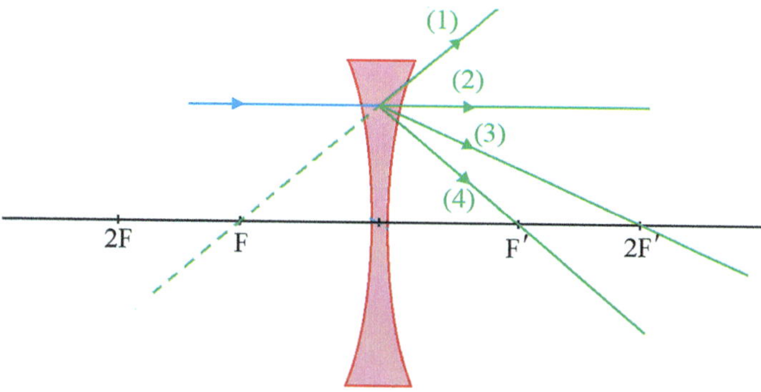

Fig. 15.13 The incident and refracted rays related to a concave lens

Problem

15.43. Which one of the choices illustrated in Fig. 15.14 shows the correct ray refracted in the concave lens?

Difficulty level ● Easy ○ Normal ○ Hard

Calculation amount ● Small ○ Normal ○ Large

1) Reflected ray (1)
2) Reflected ray (2)
3) Reflected ray (3)
4) Reflected ray (4)

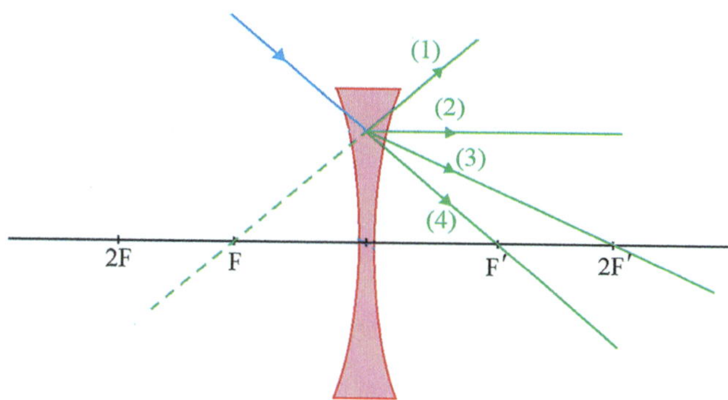

Fig. 15.14 The incident and refracted rays related to a concave lens

Problem

15.44. Which one of the choices illustrated in Fig. 15.15 shows the correct ray refracted in the concave lens?

Difficulty level ● Easy ○ Normal ○ Hard

Calculation amount ● Small ○ Normal ○ Large

1) Reflected ray (1)
2) Reflected ray (2)
3) Reflected ray (3)
4) Reflected ray (4)

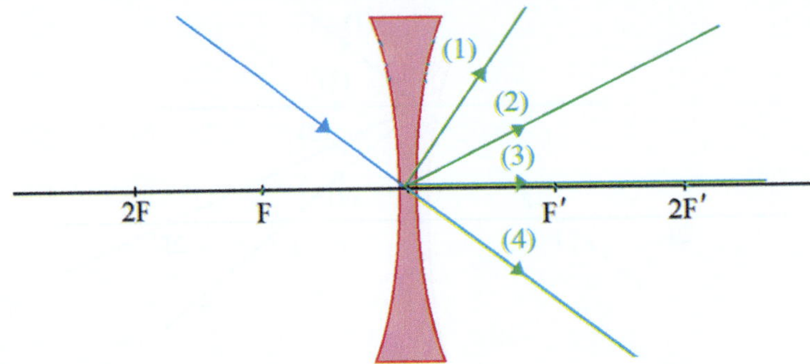

Fig. 15.15 The incident and refracted rays related to a concave lens

Problem

15.45. Which one of the following choices is wrong about the image of a concave lens.

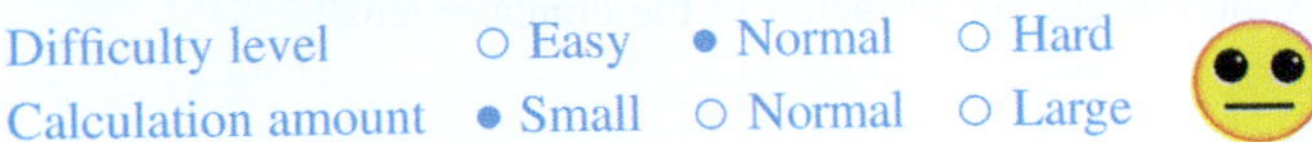

Difficulty level ○ Easy ● Normal ○ Hard

Calculation amount ● Small ○ Normal ○ Large

1) The image is always larger than the object.
2) The image is always virtual.
3) The image is always straight.
4) The image is always on the object side.

Problem

15.46. The focal distance of a concave lens is 20 *cm*. Calculate the distance of the object from the mirror for the magnification of 0.2.

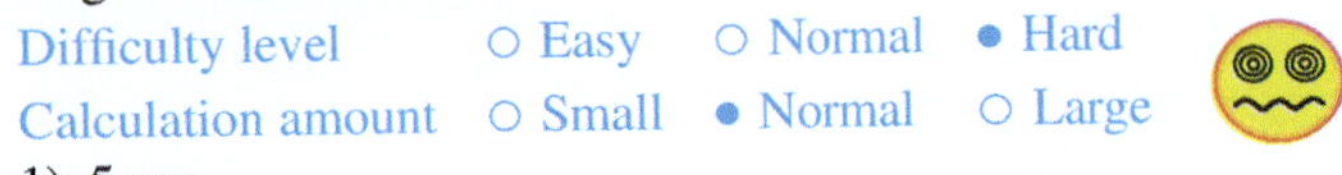

Difficulty level ○ Easy ○ Normal ● Hard

Calculation amount ○ Small ● Normal ○ Large

1) 5 *cm*
2) 80 *cm*
3) 20 *cm*
4) 40 *cm*

References

1. Rahmani-Andebili, M., General Physics I – Practice Problems, Methods, and Solutions, Springer Nature, 2025.
2. Rahmani-Andebili, M., Calculus III – Practice Problems, Methods, and Solutions, Springer Nature, 2023.
3. Rahmani-Andebili, M., Calculus II – Practice Problems, Methods, and Solutions, Springer Nature, 2023.
4. Rahmani-Andebili, M., Calculus I (2nd Ed.) – Practice Problems, Methods, and Solutions, Springer Nature, 2023.
5. Rahmani-Andebili, M., Precalculus (2nd Ed.) – Practice Problems, Methods, and Solutions, Springer Nature, 2024.

Light, Mirrors, and Lenses: Part B

16

Abstract
In this chapter, the problems of the fifteenth chapter are fully solved, in detail, step-by-step, and with different methods.

16.1 Reflection and Refraction

16.1. Based on the information given in the problem, we have [1–5]:

$$\theta_1 = 45^\circ$$

$$n_1 = 1.5$$

$$n_2 = 1.33$$

As we know, the relation between the refractive index of two environments, sine value of incident angle, and sine value of refracted angle is as follows.

$$\frac{\sin\theta_1}{\sin\theta_2} = \frac{n_2}{n_1}$$

However, first, we need to calculate the critical angle. In this condition, $\theta_2 = 90^\circ$ and $\theta_1 = \theta_c$. Thus:

$$\frac{\sin\theta_c}{\sin 90^\circ} = \frac{1.33}{1.5}$$

$$\Rightarrow \sin\theta_c = 0.878$$

$$\Rightarrow \theta_c = 62.5^\circ$$

Since $\theta_1 = 45^\circ < \theta_c = 62.5^\circ$, refraction occurs, and reflection does not happen. Hence, the refracted ray enters the water. Moreover, because water is thinner that prism, the refracted ray moves away from the normal axis. Choice (1) is the answer (Fig. 16.1).

M. Rahmani-Andebili, *General Physics II*, https://doi.org/10.1007/978-3-031-92866-6_16

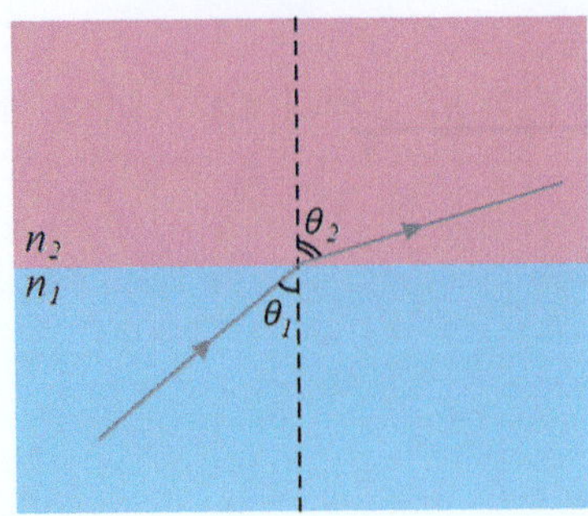

Fig. 16.1 The incident and refracted rays at the boundary of two media

Notes

In this problem, the relation below has been used.

$$\sin 90° = 1$$

16.3. Based on the information given in the problem, we have:

$$\theta_1 = 48.6°$$

$$n_1 = 1.33$$

$$n_2 = 1$$

As we know, the relation between the refractive index of two environments, sine value of incident angle, and sine value of refracted angle is as follows.

$$\frac{\sin\theta_1}{\sin\theta_2} = \frac{n_2}{n_1}$$

However, first, we need to calculate the critical angle. In this condition, $\theta_2 = 90°$ and $\theta_1 = \theta_c$. Thus:

$$\frac{\sin\theta_c}{\sin 90°} = \frac{1}{1.33}$$

$$\Rightarrow \sin\theta_c = 0.75$$

$$\Rightarrow \theta_c = 48.6°$$

As can be noticed, the angle of incident is equal to the critical angle, that is, $\theta_1 = \theta_c = 48.6°$. Therefore, the refracted ray tangentially leaves the water. In other words:

$$\theta_2 = 90°$$

Choice (4) is the answer (Fig. 16.2).

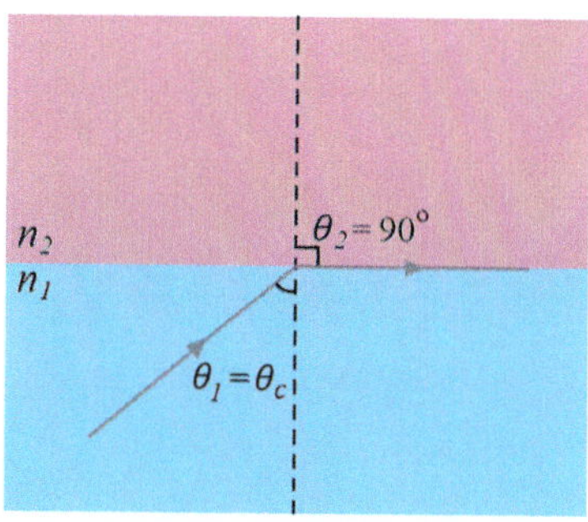

Fig. 16.2 The incident and refracted rays at the boundary of two media

Notes

In this problem, the relation below has been used.

$$\sin 90° = 1$$

16.4. Based on the information given in the problem, we have [1–5]:

$$\theta_1 = 50°$$

$$n_1 = 1.33$$

$$n_2 = 1$$

As we know, the relation between the refractive index of two environments, sine value of incident angle, and sine value of refracted angle is as follows.

$$\frac{\sin \theta_1}{\sin \theta_2} = \frac{n_2}{n_1}$$

However, first, we need to calculate the critical angle. In this condition, $\theta_2 = 90°$ and $\theta_1 = \theta_c$. Thus:

$$\frac{\sin \theta_c}{\sin 90°} = \frac{1}{1.33}$$

$$\Rightarrow \sin \theta_c = 0.75$$

$$\Rightarrow \theta_c = 48.6°$$

Since $\theta_1 = 50° > \theta_c = 48.6°$, reflection occurs. Therefore:

$$\theta_2 = \theta_1 = 50°$$

Choice (3) is the answer (Fig. 16.3).

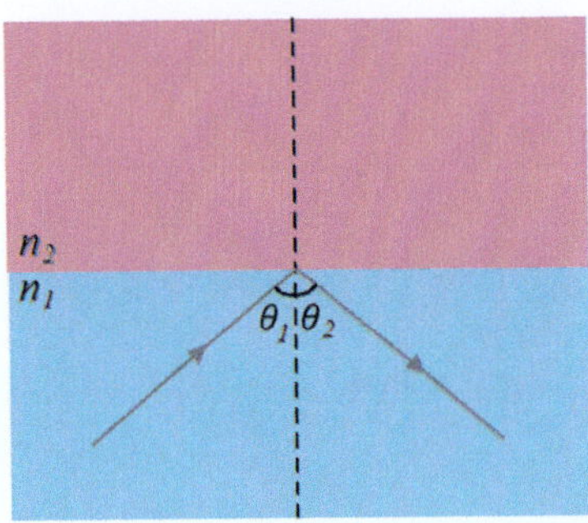

Fig. 16.3 The incident and reflected rays at the boundary of two media

Notes

In this problem, the relation below has been used.

$$\sin 90° = 1$$

16.5. Based on the information given in the problem, we have:

$$\text{Real depth} = 2\ m$$

$$n_2 = 1.33$$

The relation between the refractive index of two environments and real and apparent depths is as follows.

$$\frac{\text{Apparent depth}}{\text{Real depth}} = \frac{n_2}{n_1}$$

As we know, the refractive index of air is unity ($n_2 = 1$). Therefore:

$$\frac{\text{Apparent depth}}{2} = \frac{1}{1.33}$$

$$\Rightarrow \text{Apparent depth} = \frac{2}{1.33}$$

$$\Rightarrow \text{Apparent depth} = 1.5\ m$$

Choice (4) is the answer.

16.6. Based on the information given in the problem, we have:

$$\text{Real height} = 7.52\ m$$

$$n_2 = 1.33$$

The relation between the refractive index of two environments and real and apparent height is as follows.

$$\frac{\text{Apparent heigh}}{\text{Real height}} = \frac{n_2}{n_1}$$

As we know, the refractive index of air is unity ($n_1 = 1$). Therefore:

$$\frac{\text{Apparent heigh}}{7.52} = \frac{1.33}{1}$$

$$\Rightarrow \text{Apparent height} \approx 10\ m$$

Choice (3) is the answer.

16.2 Flat Mirror

16.7. Based on the information given in the problem and Fig. 16.4, we have:

$$\theta_i + \theta_r = 150\,^\circ$$

As we know, the incident angle and reflected angle in a flat mirror are equal. In other words:

$$\theta_i = \theta_r$$

Therefore:

$$2\theta_i = 150\,^\circ$$

$$\Rightarrow \theta_i = \theta_r = 75\,^\circ$$

Moreover, from Fig. 16.4, the following relation is noticed.

$$\theta_r + \theta = 90\,^\circ$$

Hence:

$$\theta = 15\,^\circ$$

Choice (2) is the answer.

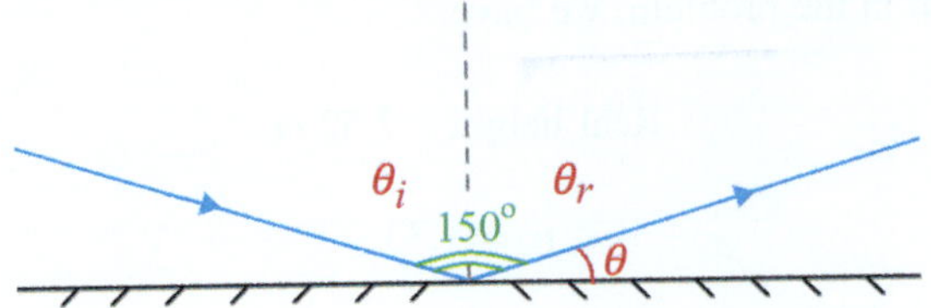

Fig. 16.4 The incident and reflected rays related to a flat mirror

---

16.8. The problem can be solved by using the following three principles.

Principle 1 A right angle is equal to 90°.

Therefore, the first incident angle can be calculated as follows.

$$90^\circ - 20^\circ = 70^\circ$$

Principle 2 The incident angle and reflected angle in a flat mirror are equal.

Hence, the first reflected angle is 70°.

Principle 1 A right angle is equal to is 90°.

Therefore, the angle between the first mirror and the first reflected ray can be calculated as follows.

$$90^\circ - 70^\circ = 20^\circ$$

Principle 3 The sum of angles of a triangle is 180°.

Thus, the angle between the second mirror and the second incident ray can be calculated as follows.

$$180^\circ - 110^\circ - 20^\circ = 50^\circ$$

Principle 1 A right angle is equal to is 90°.

Therefore, the second incident angle can be calculated as follows.

$$90^\circ - 50^\circ = 40^\circ$$

Principle 2 The incident angle and reflected angle in a flat mirror are equal.

Hence, the second reflected angle is 40°.

Principle 1 A right angle is equal to is 90°.

Therefore, θ, which is the angle between the second mirror and the second reflected ray, can be calculated as follows.

$$\theta = 90^\circ - 40^\circ$$

$$\Rightarrow \theta = 50^\circ$$

Choice (1) is the answer (Fig. 16.5).

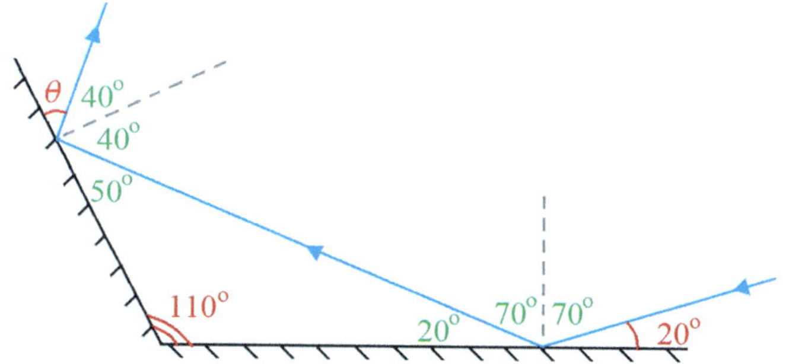

Fig. 16.5 The incident and reflected rays related to two flat mirrors

---

16.9. If an object is moved away from a flat mirror about d, its image will move away from the mirror about d. Thus, the object will move away from its image about $2d$.

Hence, for this problem, the object will move away from its image about 2 m. Choice (3) is the answer.

16.10. The number of images of an object placed between two flat mirrors, that the angle between them is α, can be calculated as follows.

$$\downarrow n = \frac{360\,^{\circ}}{\uparrow \alpha} - 1$$

As can be noticed, the number of images will decrease if α increases. Choice (2) is the answer.

16.3 Concave Mirror

16.12. A ray which is parallel to the principal axis of a concave mirror will reflect and pass from the focal point (focus). Choice (3) is the answer (Fig. 16.6).

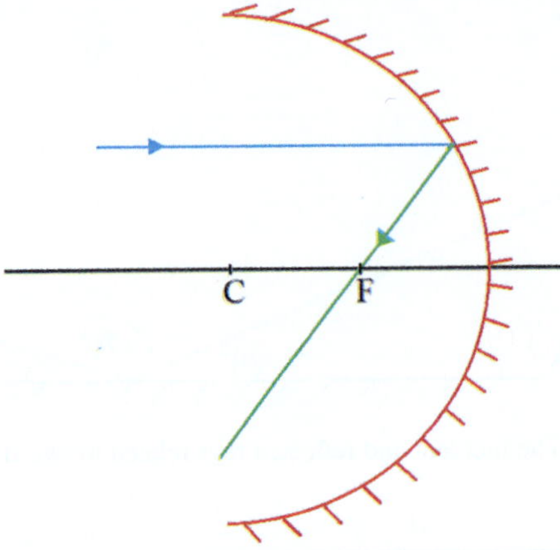

Fig. 16.6 The incident and reflected rays related to a concave mirror

16.13. A ray that passes from the focal point (focus) of a concave mirror will reflect and move parallel to the principal axis of the mirror. Choice (1) is the answer (Fig. 16.7).

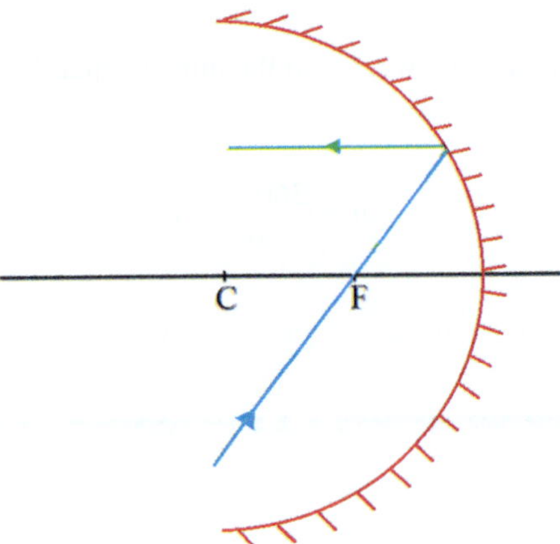

Fig. 16.7 The incident and reflected rays related to a concave mirror

16.14. A ray that passes from the center of a concave mirror will reflect on itself. Choice (2) is the answer (Fig. 16.8).

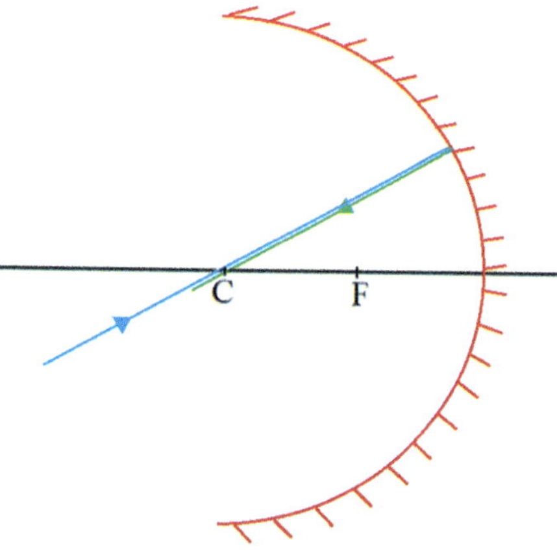

Fig. 16.8 The incident and reflected rays related to a concave mirror

16.15. As can be seen, the incident ray of the first concave mirror is parallel to its principal axis. Thus, the point p is its focal point (focus).

Moreover, the incident ray of the second concave mirror has reflected on itself. Hence, the point p is its center.

Therefore:

$$AB = f_1 + 2f_2$$

Choice (4) is the answer (Fig. 16.9).

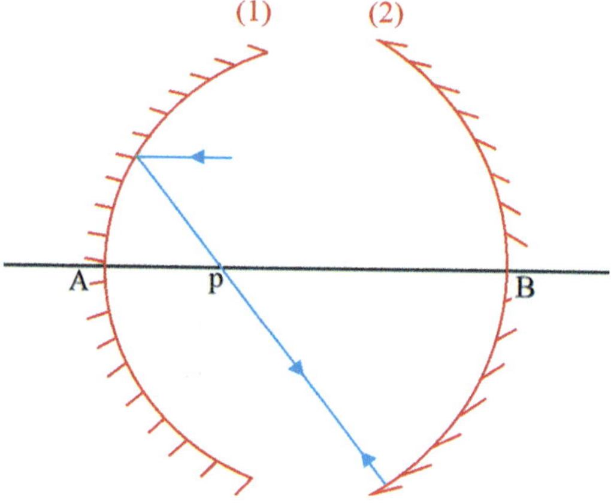

Fig. 16.9 The incident and reflected rays concerned with two parallel concave mirrors

Notes

In this problem, the relations below have been used.

For a concave mirror, we have:

$$R = 2f$$

Principal: A ray which is parallel to the principal axis of a concave mirror will reflect and pass from the focal point (focus).

Principal: A ray that passes from the center of a concave mirror will reflect on itself.

16.16. Figure 16.10 illustrates the image of an object placed in the focal distance of a concave mirror. The following facts can be extracted from Fig. 16.10.
1) The image is larger than the object.
2) The image is virtual.
3) The image is straight.
4) The image is behind the concave mirror.

Choice (1) is the answer.

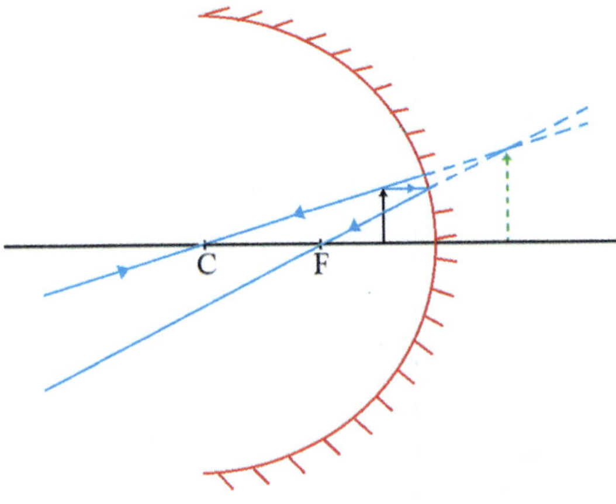

Fig. 16.10 The locations of an object and its image related to a concave mirror

---

16.17. Figure 16.11 illustrates the image of an object placed at focal point (focus) of a concave mirror. The following facts can be extracted from Fig. 16.11.
1) The image is larger than the object.
2) The image is real.
3) The image is inverted.
4) The image is at infinity.

Choice (4) is the answer.

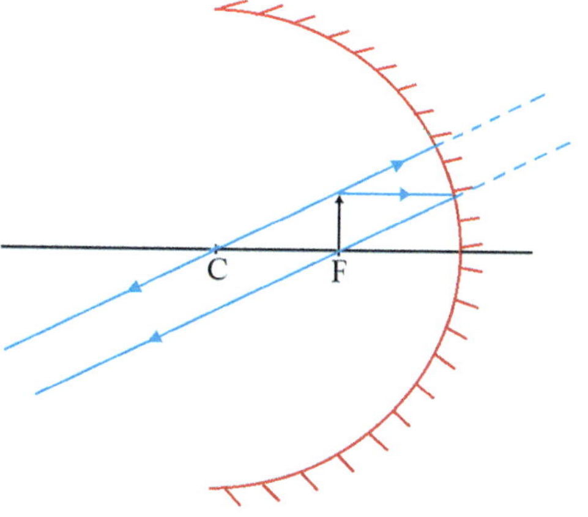

Fig. 16.11 The locations of an object and its image related to a concave mirror

16.18. Figure 16.12 illustrates the image of an object placed between the focal point (focus) and the center of a concave mirror. The following facts can be extracted from Fig. 16.12.

1) The image is larger than the object.
2) The image is real.
3) The image is inverted.
4) The image is on the object side but beyond the center of the concave mirror.

Choice (2) is the answer.

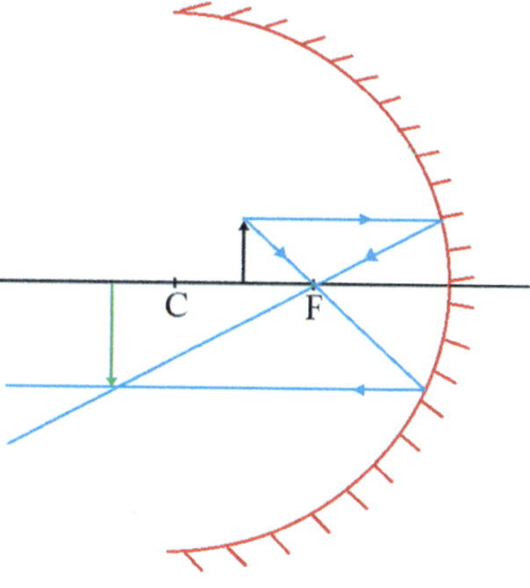

Fig. 16.12 The locations of an object and its image related to a concave mirror

16.19. Figure 16.13 illustrates the image of an object placed on the center of a concave mirror. The following facts can be extracted from Fig. 16.13.
1) The size of the image is equal to the size of the object.
2) The image is real.
3) The image is inverted.
4) The image is on the center of the concave mirror.

Choice (3) is the answer.

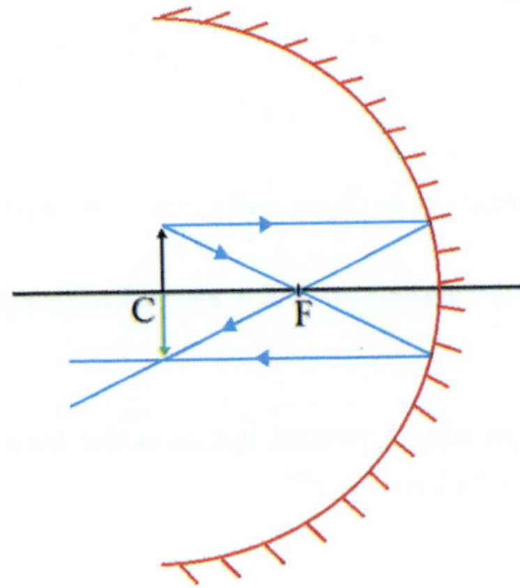

Fig. 16.13 The locations of an object and its image related to a concave mirror

---

16.20. Figure 16.14 illustrates the image of an object placed beyond the center of a concave mirror. The following facts can be extracted from Fig. 16.14.
1) The image is smaller than the object.
2) The image is real.
3) The image is inverted.
4) The image is between the focal point (focus) and the center.

Choice (4) is the answer.

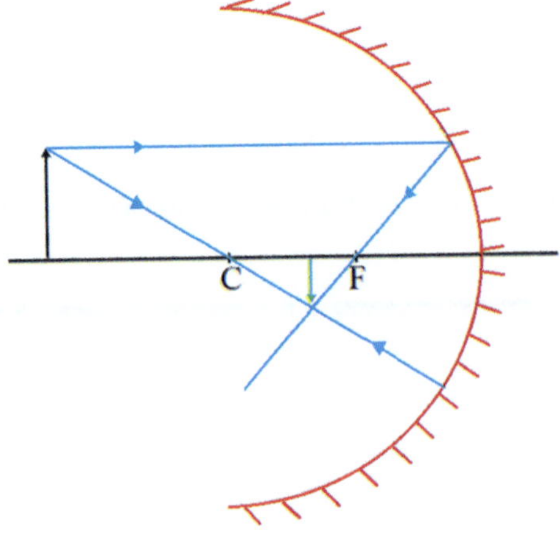

Fig. 16.14 The locations of an object and its image related to a concave mirror

---

16.21. Figure 16.15 illustrates the image of an object placed at infinity. The following facts can be extracted from Fig. 16.15.
1) The image is smaller than the object.
2) The image is real.
3) The image is inverted.
4) The image is on the focal point (focus).

Choice (1) is the answer.

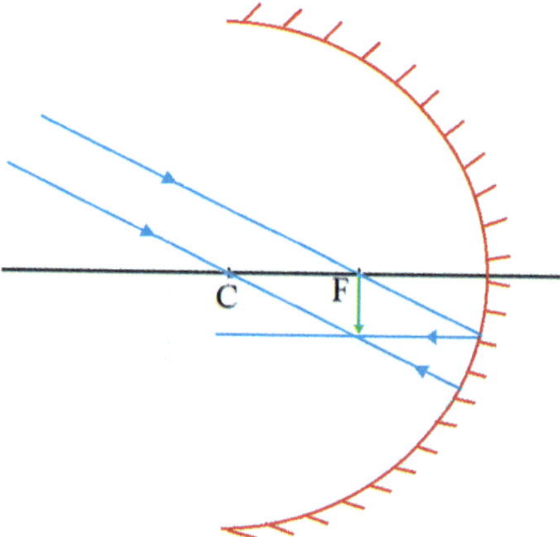

Fig. 16.15 The locations of an object and its image related to a concave mirror

---

16.22. As can be seen in Fig. 16.16, if the object is placed on the center of a concave mirror, the image is created on the center.

Moreover, as can be seen in Fig. 16.16, if the object is placed on the focal point (focus) of a concave mirror, the image is created at infinity.

Hence, it is concluded that the image will move from the center of the mirror to infinity. Choice (2) is the answer (Fig. 16.17).

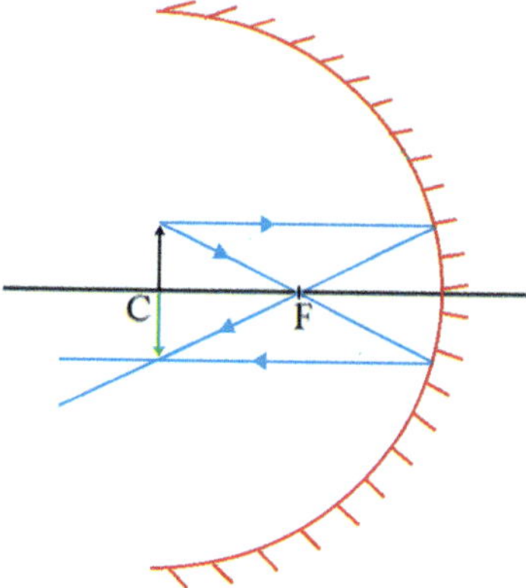

Fig. 16.16 The locations of an object and its image related to a concave mirror

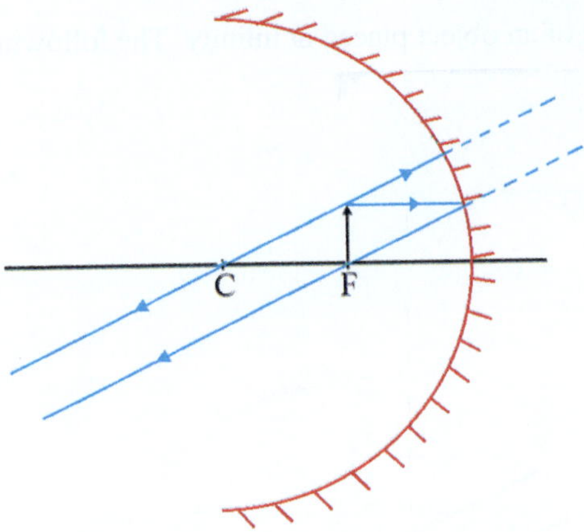

Fig. 16.17 The locations of an object and its image related to a concave mirror

16.23. Based on the information given in the problem, we have:

$$R = 40\ cm \Rightarrow f = 20\ cm$$

$$p = 30\ cm$$

As we know, for a concave mirror, the relation between the focus (f), distance of an object from the mirror (p), and distance of the image from the mirror (q) is as follows.

$$\frac{1}{p} + \frac{1}{q} = \frac{1}{f}$$

Herein, f, p, and q are positive quantities.
Moreover, the magnification of a concave mirror can be calculated as follows.

$$m = \frac{q}{p}$$

Therefore:

$$\frac{1}{30} + \frac{1}{q} = \frac{1}{20} \Rightarrow \frac{1}{q} = \frac{1}{60}$$

$$\Rightarrow q = 60\ cm$$

$$\Rightarrow m = \frac{q}{p} = \frac{60}{30}$$

$$\Rightarrow m = 2$$

Choice (1) is the answer.

Notes

In this problem, the relation below has been used.

For a concave mirror, we have:

$$R = 2f$$

16.4 Convex Mirror

16.25. A ray which is parallel to the principal axis of a convex mirror will reflect in a way that its extension will pass from the focal point (focus). Choice (1) is the answer (Fig. 16.18).

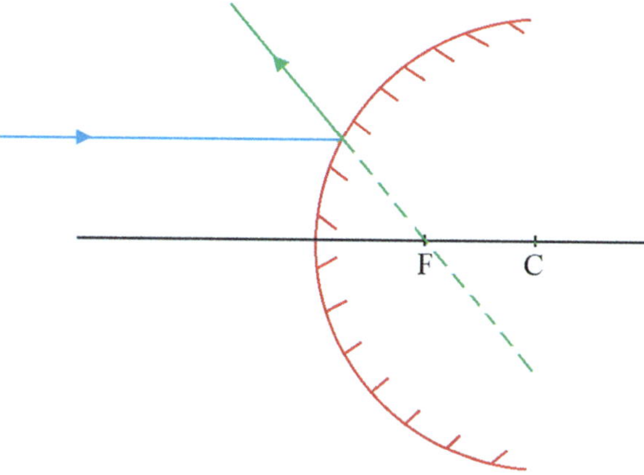

Fig. 16.18 The incident and reflected rays related to a convex mirror

16.26. A ray that its extension passes from the focal point (focus) of a convex mirror will reflect and move parallel to the principal axis of the mirror. Choice (3) is the answer (Fig. 16.19).

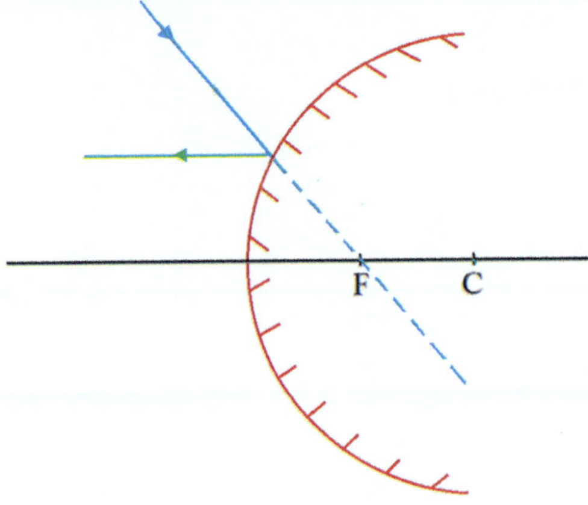

Fig. 16.19 The incident and reflected rays related to a convex mirror

---

16.27. A ray that its extension passes from the center of a convex mirror will reflect on itself. Choice (2) is the answer (Fig. 16.20).

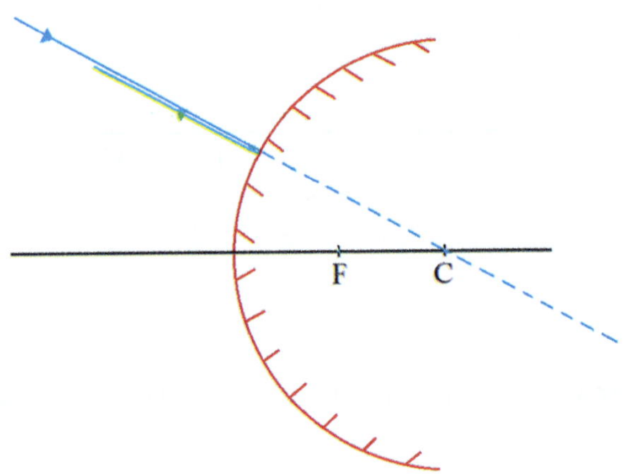

Fig. 16.20 The incident and reflected rays related to a convex mirror.

16.28. Figure 16.21 illustrates the image of an object placed on an arbitrary point in front of a convex mirror. The following facts can be extracted from Fig. 16.21.

1) The image is always smaller than the object.
2) The image is always virtual.
3) The image is always straight.
4) The image is always on the other side and on the focal distance.

Choice (4) is the answer.

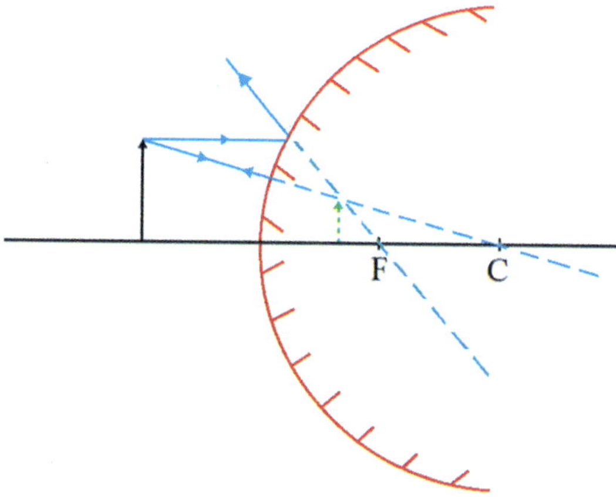

Fig. 16.21 The locations of an object and its image related to a convex mirror

16.29. Based on the information given in the problem, we have:

$$m = \frac{1}{2}$$

$$R = 28\ cm \Rightarrow f = -\frac{28}{2} = -14\ cm$$

As we know, for a convex mirror, the relation between the focus (f), distance of an object from the mirror (p), and distance of the image from the mirror (q) is as follows.

$$\frac{1}{p} + \frac{1}{q} = \frac{1}{f}$$

Herein, f and q are negative quantities.

Moreover, the magnification of a convex mirror can be calculated as follows.

$$m = -\frac{q}{p}$$

Therefore:

$$\frac{1}{2} = -\frac{q}{p} \Rightarrow q = -\frac{p}{2}$$

$$\Rightarrow \frac{1}{p} + \frac{1}{-\frac{p}{2}} = -\frac{1}{14}$$

$$\Rightarrow \frac{1}{p} - \frac{2}{p} = -\frac{1}{14}$$

$$\Rightarrow -\frac{1}{p} = -\frac{1}{14}$$

$$\Rightarrow p = 14\ cm$$

Choice (2) is the answer.

Notes

In this problem, the relation below has been used.

For a convex mirror, we have:

$$f = -\frac{R}{2}$$

16.30. Based on the information given in the problem, we have:

$$f = -20\ cm$$

$$m = 0.4$$

As we know, for a convex mirror, the relation between the focus (f), distance of an object from the mirror (p), and distance of the image from the mirror (q) is as follows.

$$\frac{1}{p} + \frac{1}{q} = \frac{1}{f}$$

Herein, f and q are negative quantities.

Moreover, the magnification of a convex mirror can be calculated as follows.

$$m = -\frac{q}{p}$$

Hence:

$$0.4 = -\frac{q}{p} \Rightarrow q = -0.4p$$

$$\Rightarrow \frac{1}{p} + \frac{1}{-0.4p} = -\frac{1}{20}$$

$$\Rightarrow \frac{1}{p} - \frac{2.5}{p} = -\frac{1}{20}$$

$$\Rightarrow \frac{-1.5}{p} = -\frac{1}{20}$$

$$\Rightarrow p = 30\ cm$$

Choice (4) is the answer.

16.5 Convex Lens

16.31. A ray which is parallel to the principal axis of a convex lens will refract and pass from the focal point (focus). Choice (3) is the answer (Fig. 16.22).

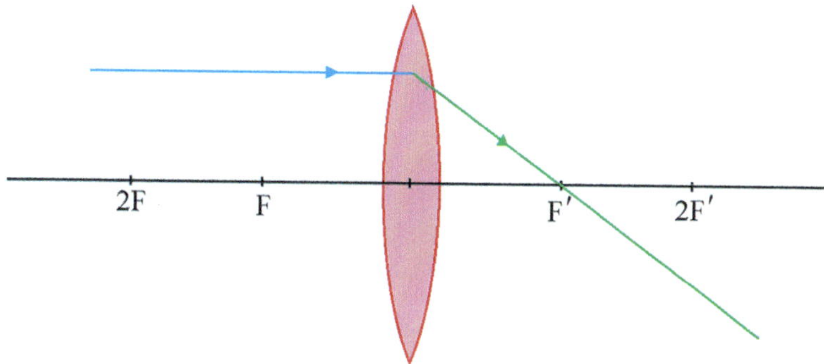

Fig. 16.22 The incident and refracted rays related to a convex lens

16.32. A ray that passes from the focal point (focus) of a convex lens will refract and move parallel to the principal axis of the lens. Choice (2) is the answer (Fig. 16.23).

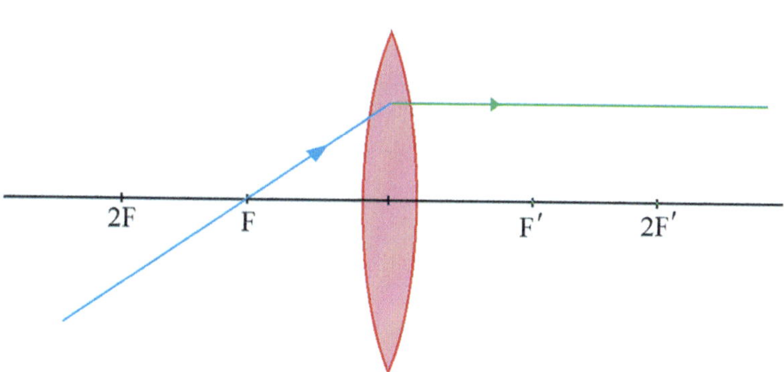

Fig. 16.23 The incident and refracted rays related to a convex lens

16.33. A ray that passes from the center of the curvature of a convex lens will not refract. Choice (2) is the answer (Fig. 16.24).

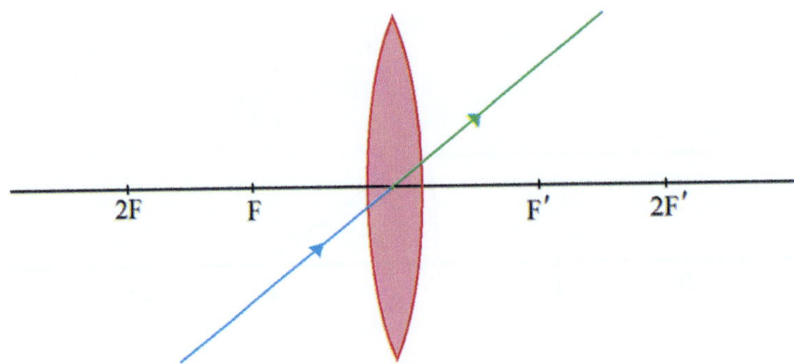

Fig. 16.24 The incident and refracted rays related to a convex lens

16.34. Figure 16.25 illustrates the image of an object placed in the focal distance of a convex lens. The following facts can be extracted from Fig. 16.25.

1) The image is larger than the object.
2) The image is virtual.
3) The image is straight.
4) The image is on the object side.

Choice (2) is the answer.

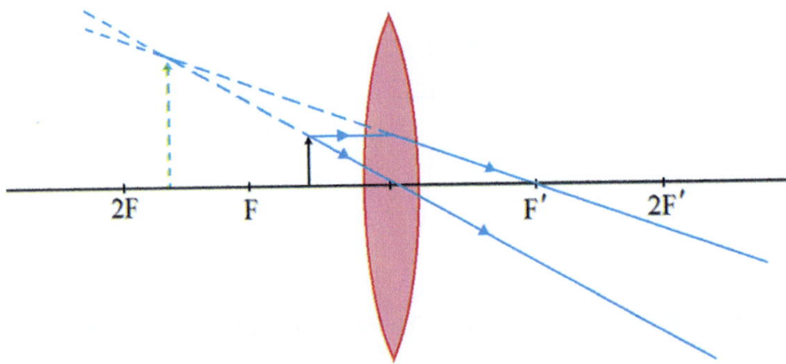

Fig. 16.25 The locations of an object and its image related to a convex lens

16.35. Figure 16.26 illustrates the image of an object placed at the focal point (focus). The following facts can be extracted from Fig. 16.26.
1) The image is larger than the object.
2) The image is real.
3) The image is inverted.
4) The image is at infinity.

Choice (2) is the answer.

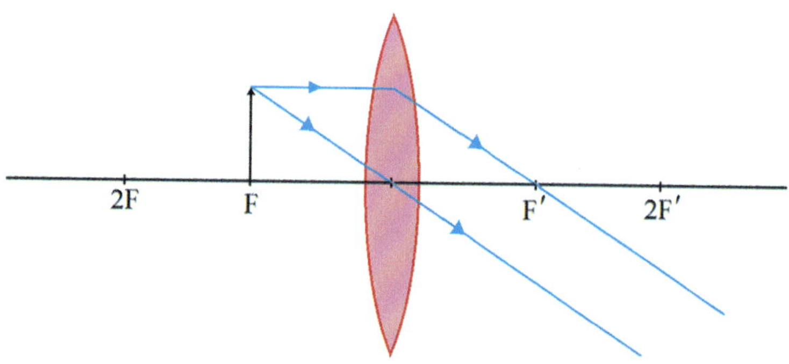

Fig. 16.26 The locations of an object and its image related to a convex lens

16.36. Figure 16.27 illustrates the image of an object placed between the focal point (focus) and the center of the curvature of the lens. The following facts can be extracted from Fig. 16.27.
1) The image is larger than the object.
2) The image is real.
3) The image is inverted.
4) The image is on the other side of the lens and beyond the center of the curvature of the lens.

Choice (3) is the answer.

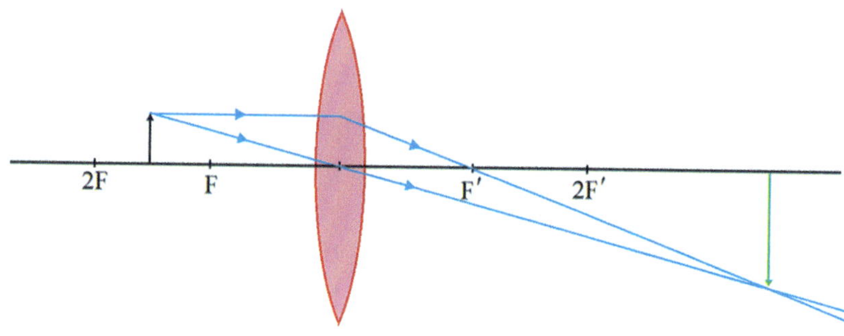

Fig. 16.27 The locations of an object and its image related to a convex lens

16.37. Figure 16.28 illustrates the image of an object placed on the center of the curvature of a convex lens. The following facts can be extracted from Fig. 16.28.

1) The size of the image is equal to the size of the object.
2) The image is real.
3) The image is inverted.
4) The image is on the other side, and is created on the center of the curvature of the lens.

Choice (4) is the answer.

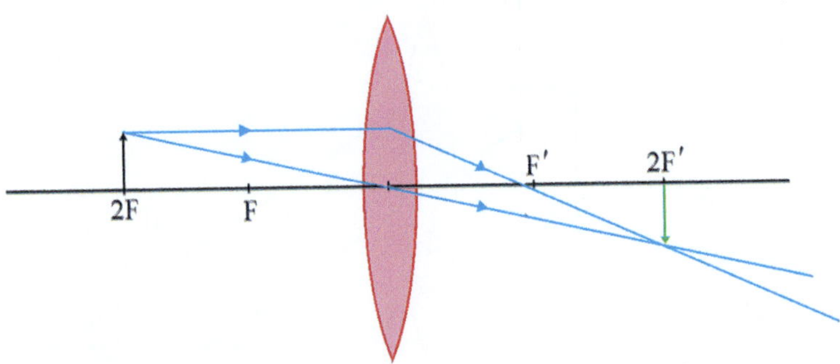

Fig. 16.28 The locations of an object and its image related to a convex lens

16.38. Figure 16.29 illustrates the image of an object placed beyond the center of the curvature of a convex lens. The following facts can be extracted from Fig. 16.29.

1) The image is smaller than the object.
2) The image is real.
3) The image is inverted.
4) The image is on the other side of the lens and between the focal point and the center of the curvature of the lens.

Choice (1) is the answer.

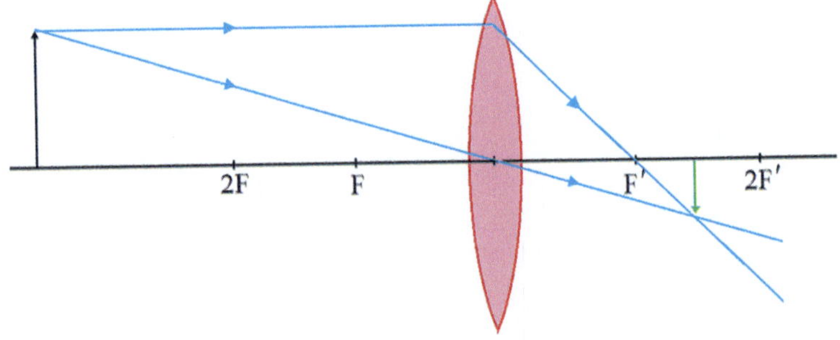

Fig. 16.29 The locations of an object and its image related to a convex lens

16.39. Figure 16.30 illustrates the image of an object placed at infinity. The following facts can be extracted from Fig. 16.30.
1) The image is smaller than the object.
2) The image is real.
3) The image is inverted.
4) The image is on the focal point (focus).

Choice (4) is the answer.

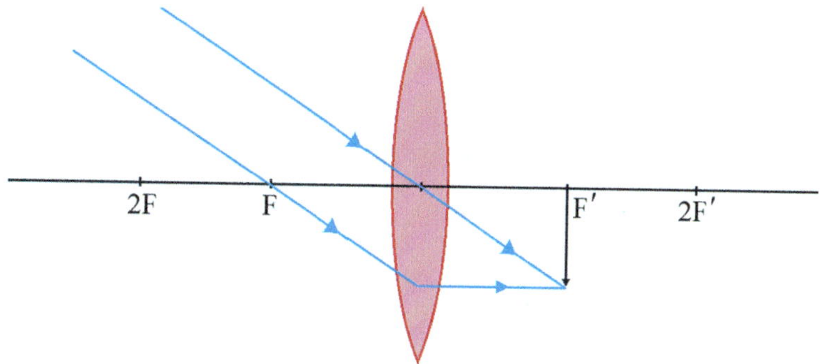

Fig. 16.30 The locations of an object and its image related to a convex lens

16.40. Based on the information given in the problem, we have:

$$f = 20\ cm$$

$$p = 30\ cm$$

As we know, for a convex lens, the relation between the focus (f), distance of an object from the lens (p), and distance of the image from the lens (q) is as follows.

$$\frac{1}{p} + \frac{1}{q} = \frac{1}{f}$$

Herein, f, p, and q are positive quantities.

Moreover, the magnification of a convex lens (m) can be calculated as follows.

$$m = \frac{q}{p}$$

Thus:

$$\frac{1}{30} + \frac{1}{q} = \frac{1}{20} \Rightarrow \frac{1}{q} = \frac{1}{60}$$

$$\Rightarrow q = 60\ cm$$

$$m = \frac{q}{p} = \frac{60}{30}$$

$$\Rightarrow m = 2$$

Choice (1) is the answer.

16.6 Concave Lens

16.42. A ray which is parallel to the principal axis of a concave lens will reflect in a way that its extension will pass from the focal point (focus). Choice (1) is the answer (Fig. 16.31).

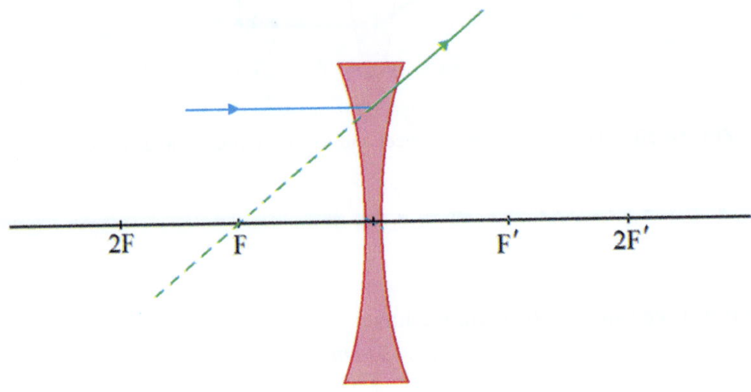

Fig. 16.31 The incident and refracted rays related to a concave lens

16.43. A ray that its extension passes from the focal point (focus) of a concave lens will reflect and move parallel to the principal axis of the lens. Choice (2) is the answer (Fig. 16.32).

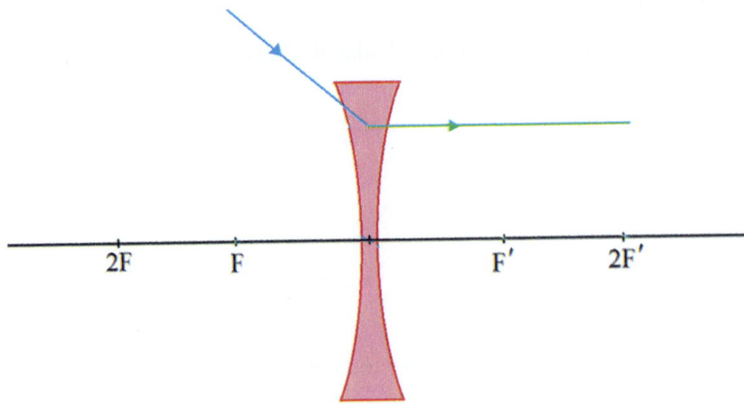

Fig. 16.32 The incident and refracted rays related to a concave lens

16.44. A ray that passes from the center of a concave lens will not reflect. Choice (4) is the answer (Fig. 16.33).

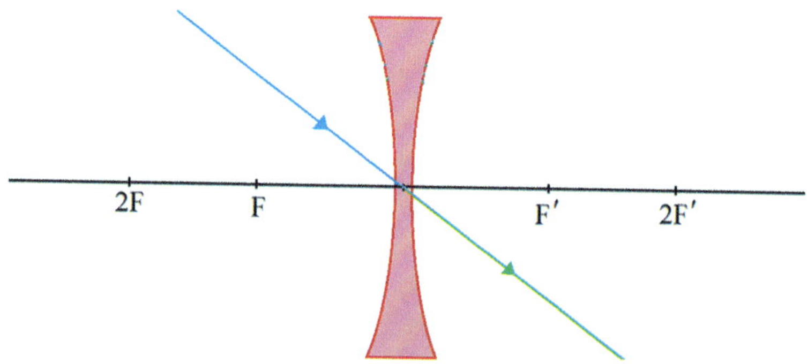

Fig. 16.33 The incident and refracted rays related to a concave lens

---

16.45. Figure 16.34 illustrates the image of an object placed on an arbitrary point in front of a concave lens. The following facts can be extracted from Fig. 16.34.

1) The image is always smaller than the object.
2) The image is always virtual.
3) The image is always straight.
4) The image is always on the object side.

Choice (1) is the answer.

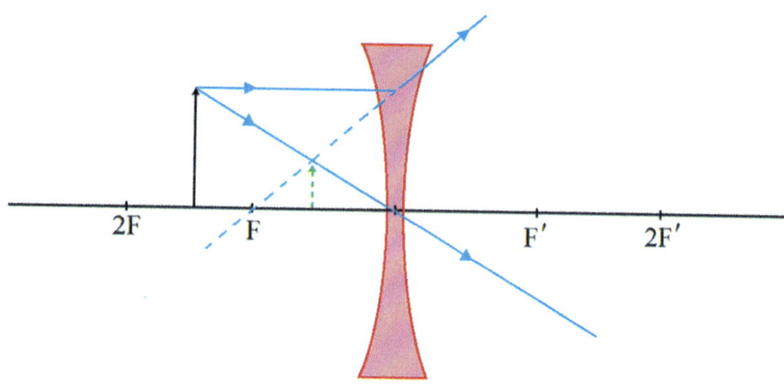

Fig. 16.34 The locations of an object and its image related to a concave lens

---

16.46. Based on the information given in the problem, we have:

$$f = -20\ cm$$

$$m = 0.2$$

As we know, for a concave lens, the relation between the focus (f), distance of an object from the mirror (p), and distance of the image from the mirror (q) is as follows.

$$\frac{1}{p} + \frac{1}{q} = \frac{1}{f}$$

Herein, f and q are negative quantities.

Moreover, the magnification of a concave lens can be calculated as follows.

$$m = -\frac{q}{p}$$

Hence:

$$0.2 = -\frac{q}{p} \Rightarrow q = -0.2p$$

$$\Rightarrow \frac{1}{p} + \frac{1}{-0.2p} = -\frac{1}{20}$$

$$\Rightarrow \frac{1}{p} - \frac{5}{p} = -\frac{1}{20}$$

$$\Rightarrow \frac{-4}{p} = -\frac{1}{20}$$

$$\Rightarrow p = 80\ cm$$

Choice (2) is the answer.

References

1. Rahmani-Andebili, M., General Physics I – Practice Problems, Methods, and Solutions, Springer Nature, 2025.
2. Rahmani-Andebili, M., Calculus III – Practice Problems, Methods, and Solutions, Springer Nature, 2023.
3. Rahmani-Andebili, M., Calculus II – Practice Problems, Methods, and Solutions, Springer Nature, 2023.
4. Rahmani-Andebili, M., Calculus I (2nd Ed.) – Practice Problems, Methods, and Solutions, Springer Nature, 2023.
5. Rahmani-Andebili, M., Precalculus (2nd Ed.) – Practice Problems, Methods, and Solutions, Springer Nature, 2024.

Index

M. Rahmani-Andebili, *General Physics II*, https://doi.org/10.1007/978-3-031-92866-6

The manufacturer's authorised representative in the EU is Springer Nature Customer Service Centre GmbH, Europaplatz 3, 69115 Heidelberg, Germany. If you have any concerns regarding our products, please contact ProductSafety@springernature.com

Printed and bound by CPI Group (UK) Ltd, Croydon, CR0 4YY
07/07/2026
02160934-0001